How Wireless Works

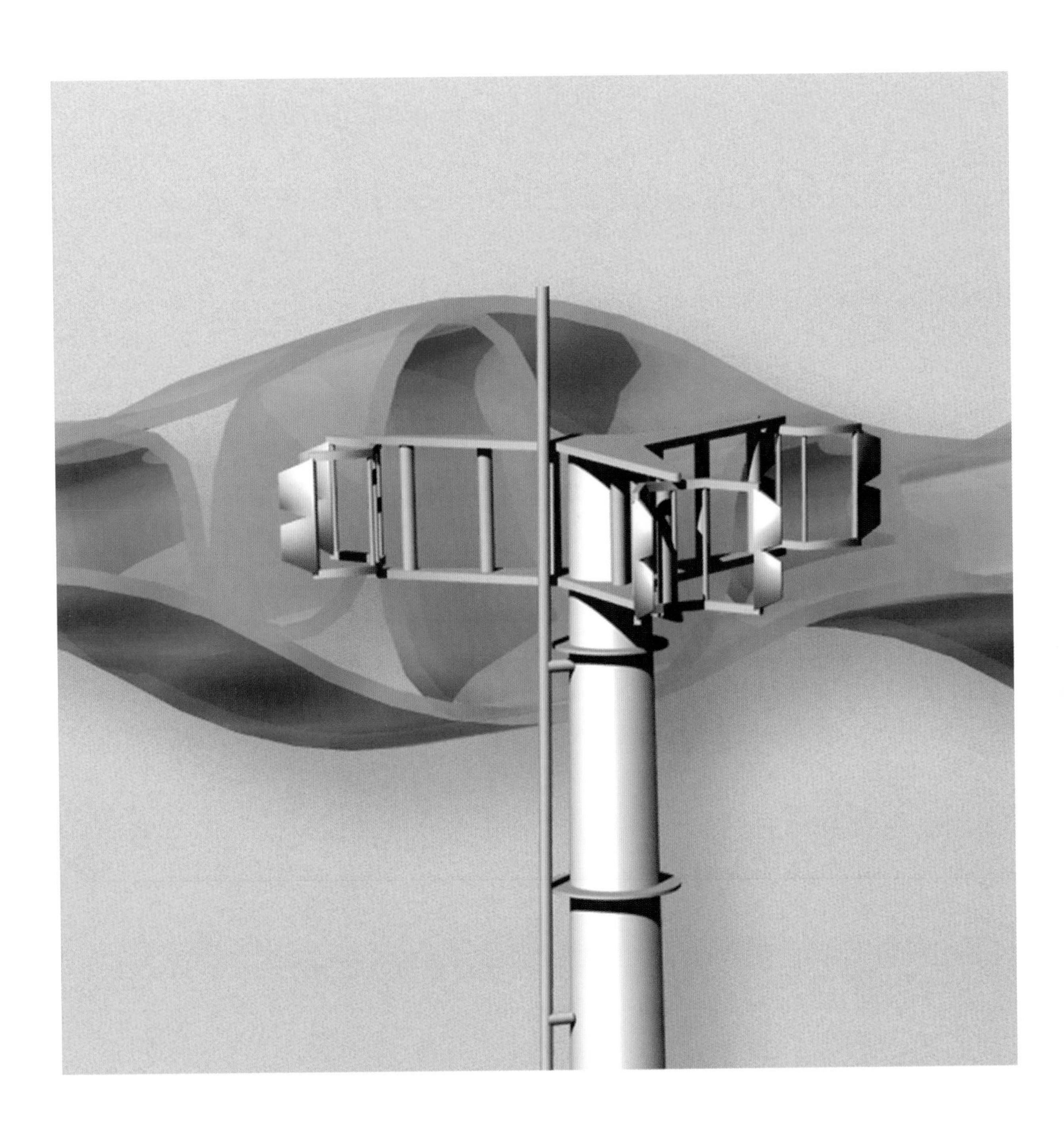

How Wireless Works

Preston Gralla

201 W. 103rd Street
Indianapolis, IN 46290

How Wireless Works

Associate Publisher	Greg Wiegand
Acquisitions Editor	Stephanie McComb
Development Editor	Nicholas J. Goetz
Managing Editor	Thomas F. Hayes
Project Editor	Tonya Simpson
Indexer	Sandra Henselmeier
Proofreader	Maribeth Echard
Technical Editor	Don Marsh
Illustrator	D&G Limited
Team Coordinator	Sharry Lee Gregory
Interior Designers	Anne Jones and Dan Armstrong
Cover Designer	Alan Clements
Page Layout	Gloria Schurick

International Standard Book Number: 0-7897-2487-1

Library of Congress Catalog Card Number: 00-108245

Printed in the United States of America

First Printing: September 2001

04 03 02 01 4 3 2 1

Acknowledgments

AS with all books, this was a collaboration among many people, and although my name is on the cover, thanks have to go to them. Thanks to Greg Wiegand and Stephanie McComb for trusting me with the book; and to Nick Goetz for helping shape it, working with the illustrators and in general for pulling together all the elements in a book as complex as this one.

Thanks also to technical editor Don Marsh, and illustrators at D&G Limited: Erikk D. Lee and Colin King. And there are many people at Que who deserve recognition including Tonya Simpson, Gloria Schurick, Maribeth Echard, and Sandy Henselmeier.

Without the aid of the many people and companies whom I interviewed for this book, it wouldn't exist. The people at Logitech, ShareWave, and Netgear were especially helpful in providing much-needed information.

Finally, the biggest thanks as always to my wife, Lydia, and my kids, Gabe and Mia. Doing this book so monopolized my life at times, that probably the easiest way for them to reach me would have been through the ultimate wireless technology—the Vulcan mind meld.

Tell Us What You Think!

As the reader of this book, *you* are our most important critic and commentator. We value your opinion and want to know what we're doing right, what we could do better, what areas you'd like to see us publish in, and any other words of wisdom you're willing to pass our way.

As an associate publisher for Que, I welcome your comments. You can fax, e-mail, or write me directly to let me know what you did or didn't like about this book—as well as what we can do to make our books stronger.

Please note that I cannot help you with technical problems related to the topic of this book, and that due to the high volume of mail I receive, I might not be able to reply to every message.

When you write, please be sure to include this book's title and author as well as your name and phone or fax number. I will carefully review your comments and share them with the author and editors who worked on the book.

Fax: 317-581-4666

E-Mail: feedback@quepublishing.com

Mail: Greg Wiegand
Que
201 West 103rd Street
Indianapolis, IN 46290 USA

Introduction

YOU take it for granted: You pick up a cell phone, make a call, hang up, and then go about your business. You tune in your radio to a baseball game occurring on the other side of the continent. You watch a war taking place live on the other side of the planet. You page the plumber because your hot water heater has broken and is flooding your basement. You turn off your car alarm by remote control.

Welcome to the wireless world. Just about every aspect of your daily life is touched in one way or another by wireless technology, by the sending of signals and information through the air. It suffuses our life, and yet, little more than 100 years ago, people didn't even realize that waves could carry information.

Probably the closest thing the modern world has to magic is wireless technology. Invisibility; things appearing out of thin air; communicating across the street, across town, across the continent or the world—it has all the earmarks of magic.

Although we all use wireless technologies many times a day, most of us probably have only the vaguest idea of how the technologies actually work. Perhaps we have some notion that some kind of waves carry information in some way. We've probably heard the term modulation or amplification, or cell or base station. But as for the details…they seem to escape us.

This book is dedicated to demystifying how wireless technologies work. As you'll see, they're not that mysterious or difficult to understand. The book covers everything from the basics of the electromagnetic spectrum to how next-generation wireless technologies work and will change our lives. Whether you don't have a clue about how wireless works or consider yourself something of a cellular maven, there's something here for you to learn. Making it all the easier for you is that it's all explained in easy-to-follow, glorious, full-color illustrations. So, no matter how complex and intricate the topic, you'll find it easy to follow.

Part 1, "Understanding Wireless's Basic Technologies," introduces you to the most basic principles of how wireless works. You'll learn about basic wireless concepts, see a timeline of how wireless technologies have developed, and see the many ways in which wireless technologies are used in our everyday lives. This section of the book also explains the electromagnetic spectrum, details what the radio frequency spectrum is, shows you how electromagnetic waves are created, and describes how data is transmitted by them. You'll be introduced to a basic wireless network and learn about amplitude modulation and frequency modulation—the two basic ways in which data is put onto RF waves. And the section covers the basic hardware as well—you'll see how antennas, transmitters, and receivers work.

Part 2, "How Radio and Television Work," details how those two common broadcasting media work. You'll see how radio broadcasts are created and transmitted, and then can be tuned in by your radio and played. You'll learn about low-power FM broadcasting—a technology approved by the Federal Communications Commission that allows nonprofit groups to create their own radio stations and broadcast in a small area, such as a neighborhood or town. And you'll learn about the newest technology to hit radio: subscription satellite broadcasts. For a fee of about $10 a month, you'll be able to listen to hundreds of high-quality broadcasts. And because the broadcasts are delivered via satellite, you'll be able to listen to those radio stations wherever you are. This part also shows how the magic of TV broadcasting works, how the signal is created, processed, and sent through the air. You'll also learn about high-definition TV, the next big thing in television, and how satellite dishes work.

Part 3, "How Cellular Telephones and Pagers Work," shows you the intricacies of the most popular cellular communications technologies. You'll look inside cell phones and pagers so that you can see how the devices do their processing, and what the electronic components do. You'll learn about cells and base stations and how they work together with your phone so you can be located to receive phone calls, and easily make them when you need. This section also explains the differences between a whole alphabet soup of cell-phone technologies: GSM, PCS, TDMA, CDMA, and more. And you'll find out about the differences between digital and analog cell phones as well. By the time you're finished with this part, there will be hardly a thing you won't understand about cell phones and pagers.

Part 4, "Understanding Wireless Networks," looks at how wireless technology is used to connect computers and allow them to communicate with one another. There's no doubt that, just as telephones are increasingly becoming wireless, the same thing will happen to computer networks. You'll see how home wireless networks function, as well as wireless networks in large corporations. Bluetooth and the 802.11 standard are the two primary ways that computers (and increasingly, other devices) communicate wirelessly. You'll see exactly how those standards work and how they'll be used in the future.

Part 5, "The Wireless Internet," details the convergence of the two great technologies of our time, the world-spanning Internet and wireless communications. Although today most people access the Internet using wired computers, that won't be the case in the future. In fact, some people believe that in the not-too-distant future, the Internet will be trafficked by more wireless devices than by wired computers. This section of the book shows the basics of the Internet, and then shows how cell phones and personal digital assistants (PDAs) get onto the Internet. You'll learn about a variety of technologies, including the Wireless Application Protocol (WAP), the Wireless Markup Language (WML), WMLScript, Web clipping, XML, Voice XML, and many others. You'll also see how wireless PDAs access the Internet, how wireless keyboards and mice work, and how computers can print without wires. And you'll learn about the most advanced use of Internet-enabled wireless technology, the i-mode cell phones that had their start in Japan.

The last section of the book, Part 6, "Applying Wireless Technology: mCommerce, Security, Business Use, and Beyond," shows you the many uses to which wireless technologies have been put. You'll learn how cell phones will be used for commerce, how corporations incorporate wireless technologies into their computer systems, and about all the dangers to your privacy and security posed by wireless technologies, such as wireless viruses and cell-phone snoopers. This section also describes the next generation of wireless technology, so-called 3G (for third-generation) technologies. And you'll see some of the more amazing uses of wireless technology, such as how it's used in satellite transmissions and satellite phones, and how space exploration satellites use wireless technology to communicate with Earth.

So, come along to learn about the invisible world of communications all around us. As you'll see in this book, it's not really magic—and in learning about it, you'll see that the reality of how it works is more amazing than any magic could ever be.

P A R T

I

Understanding Wireless's Basic Technologies

AS I've said in this book's introduction, wireless technologies are the closest thing that the modern world has to magic. The capability to make things appear at long distance, traveling through the invisible ether—wireless technology has all the earmarks of prestidigitation.

As we all know, there is no such thing as magic. And the truth is, what appears to be magic is just the culmination of a series of basic laws of nature and basic applied technologies that makes remarkable things possible.

In this section of the book, we're going to look at those wireless basics that make it all possible—that allow everything from TV and radio waves to travel through the air and allow your TV and radio to play the signals; that allow cellular communications across the planet; that allow interstellar communications; and more.

Chapter 1, "Welcome to the World of Wireless," takes a "big-picture" approach to understanding wireless technologies. It starts off by looking at a timeline of wireless technologies. We'll peer into the previous centuries to see when the electromagnetic spectrum and radio waves were first discovered; we'll see when early wireless technologies, such as radio, were conceived; we'll look at when the now-mature technology of TV got its start and reached fruition; and we'll see a timeline of modern wireless technologies, such as cellular telephones and beyond. In this chapter, we'll also get our first look at basic wireless concepts, such as modulation, cells, transmitters, and receivers. And we'll see how wireless technologies are used in our everyday lives.

Chapter 2, "What Is the Electromagnetic Spectrum?" introduces the most basic concept of all wireless technologies—the electromagnetic spectrum. The spectrum is made up of energy waves that do everything from let us see the world, to cook our dinner, to let us peer into the body with X-rays, and, of course, to communicate using wireless technology. We'll take an in-depth look at the spectrum and how it works, as well as the specific part of the electromagnetic spectrum used for communications—what are called the radio frequencies, or *RF*. And of particular importance, we'll see how electromagnetic waves are created.

Chapter 3, "How a Basic Wireless System Works," starts getting down to the nitty-gritty. We'll overview a basic wireless network. We'll see how information is modulated onto RF waves, is sent through a transmitter, travels through the air, and then is demodulated at the receiving end so it can be understood.

Chapter 4, "How Amplitude Modulation (AM) and Frequency Modulation (FM) Work," looks inside the mysteries of modulation. All information that needs to be transmitted wirelessly, whether it's your voice, TV signals, or digital data, needs to be transferred onto carrier waves that send that information through the air. The information is put onto the waves in a process called *modulation*. There are two ways that data can be modulated onto a wave—*amplitude modulation* (AM) and *frequency modulation* (FM). In this chapter, we'll see how each of those techniques works and learn the pros and cons of using each different kind.

In Chapter 5, "How Data Rides on Wireless Waves," we'll learn more about how information travels on carrier waves. We'll look at what happens to a signal after it has been modulated—we'll see, for example, how it's processed by devices such as signal processors so that it can be sent through the air most effectively. We'll also learn about things such as signal gain and how interference affects RF waves. The chapter covers some very basic and important information—understanding the difference between sending analog information over RF waves and sending digital information over RF waves.

Chapter 6, "How Antennas, Transmitters, and Receivers Work," covers the most basic hardware in any wireless system—antennas, transmitters, and receivers. There's no way that signals can get from point A to point B unless this hardware is there to do it. You'll learn not only the inner workings of these devices, but also gain an understanding of the importance of various antenna, transmitter, and receiver designs.

So, whether you're a wireless maven or just trying to understand how wireless technology works, you'll find a lot in this first part to help you understand the wireless world. As you'll see, wireless technologies are, in a way, magic—although magic of the most practical sort.

CHAPTER

1

Welcome to the World of Wireless

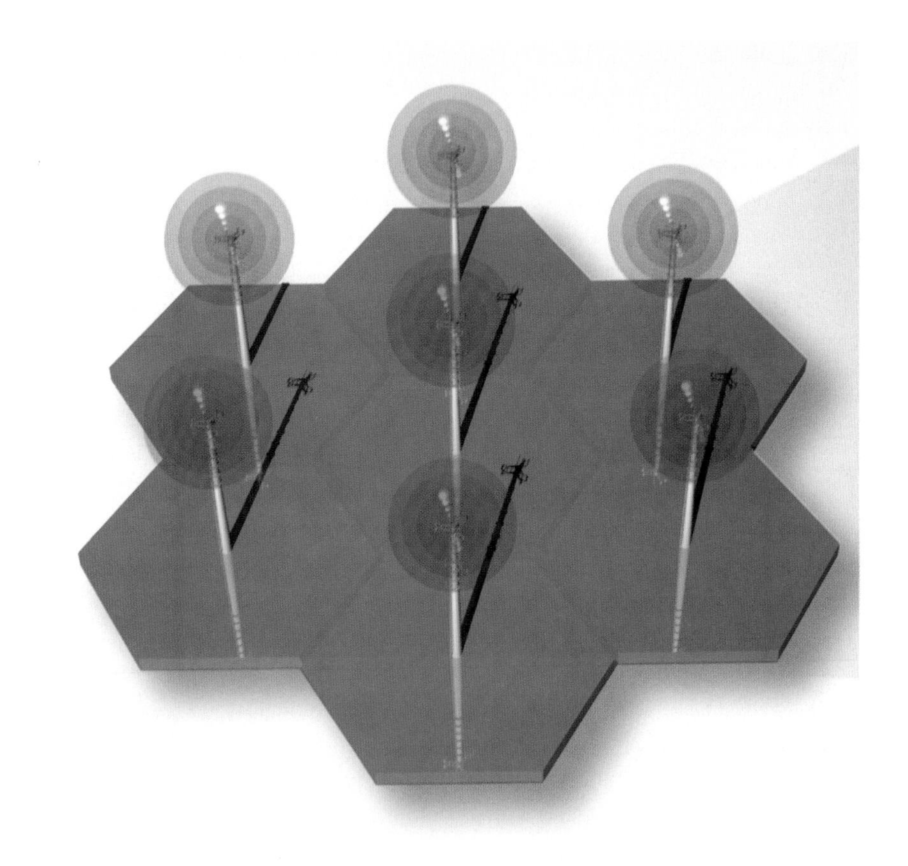

A little more than 100 years ago, an Italian physicist and inventor named Guglielmo Marconi was the first person to successfully transmit information over radio waves, and the world has never been the same since.

These days, it's chic to talk about the computer revolution or the Internet revolution as the driving force behind changes in the way we live and work, but in fact, the greatest and most far-reaching revolution of the last 100 years or so has been neither of them—it's been wireless technology, the capability to send information up to thousands of miles invisibly through the air.

Without wireless transmissions there would be no broadcast mass media. No radio. No television. No instant communications via satellites. The world has become a global village, in large part because of wireless technology—and in fact, in many of the poorer countries on earth it's easier to communicate using cell phones than over traditional telephones (called *landlines*) because of the immense cost involved in stringing telephone wires over vast distances.

Despite all the advances in wireless communications—the cell phones, the pagers, satellite transmissions, the transmission of digital data—wireless technology works to a great extent the same way today as it did back in the days of Marconi.

All wireless transmissions—whether Morse code in the days of Marconi or digital data today—are able to piggyback information onto invisible waves. These waves are part of the *electromagnetic spectrum*—energy waves that include visible light, X-rays, ultraviolet light, microwaves, and many other kinds of waves. The portion of the electromagnetic spectrum that can be used to transmit information is called the *radio frequency (RF)*. RF is used to transmit all kinds of data, not just radio broadcasts, so don't be confused by the term. The information piggybacked onto RF waves can be of any kind—anything from voice to television signals to computer data.

The information is piggybacked onto the waves using a device called a *modulator*. It's then transmitted through the air. Sometimes it's sent to a device a foot away from the transmitter; other times it's broadcast to a wide audience hundreds or thousands of miles away; and yet other times it might be sent to a cell phone tower a mile away, where it then is sent by that tower somewhere closer to its destination. But in all cases, when it reaches its destination, the information is taken from the wave in a process called *demodulation*. And that process of modulating information onto a wave, transmitting the wave, and then receiving the wave and demodulating information from it, is the heart of all wireless technology.

Whether you're doing something as simple as changing the channel on your TV or accomplishing a task a bit more complex, such as sending and receiving e-mail on your cell phone, realize that you're using technology that's older than a century, and as up to date as today's news.

Understanding Basic Wireless Concepts

ON AIR
ON AIR

Modulator

Data

Modulation Information is piggybacked onto RF waves by a process called modulation. When the wave is received, it must be demodulated to extract the information out of it.

Data Information of many kinds can be sent wirelessly—anything from radio and TV signals, to the human voice, to computer data. The information is sent by piggybacking it onto *radio waves*—electromagnetic energy that occupies a specific portion of the electromagnetic spectrum—the radio frequency (RF) portion. All kinds of data can be sent using RF, not just AM and FM radio signals. For more information about the electromagnetic spectrum and RF, turn to Chapter 2, "What Is the Electromagnetic Spectrum?"

Cells The name *cellular telephone* comes from the concept of the *cell*, which divides an area into several small cells. When a cell phone sends or receives calls, it communicates within that cell, so that no long-distance communication needs to take place. The information then is sent from that cell on to its eventual destination.

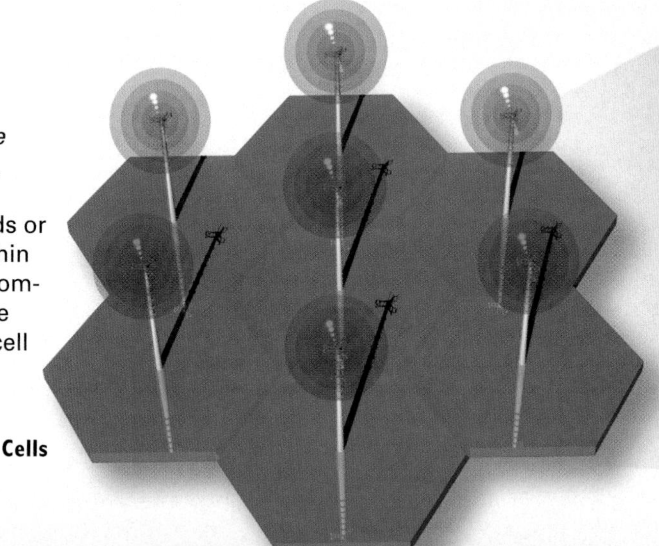

Cells

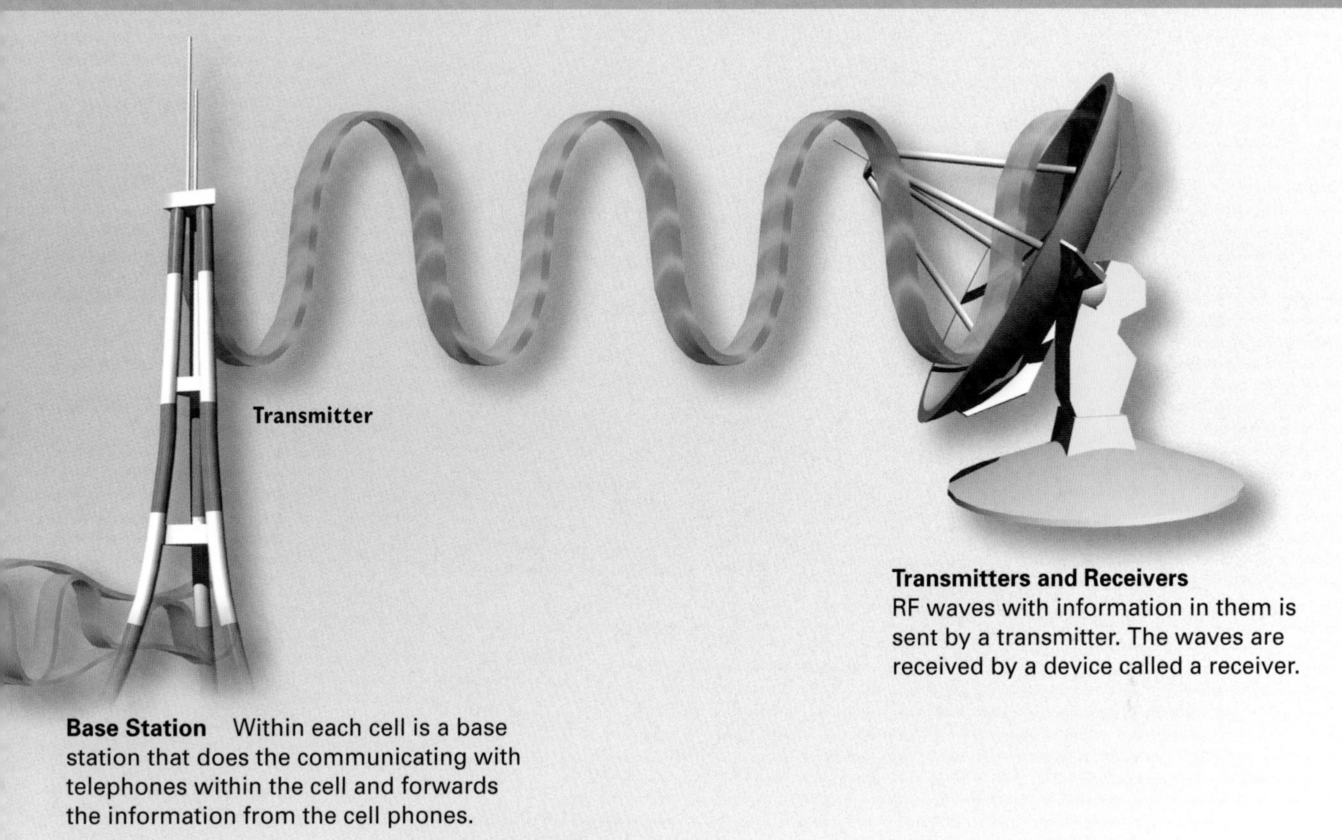

Transmitters and Receivers
RF waves with information in them is sent by a transmitter. The waves are received by a device called a receiver.

Base Station Within each cell is a base station that does the communicating with telephones within the cell and forwards the information from the cell phones.

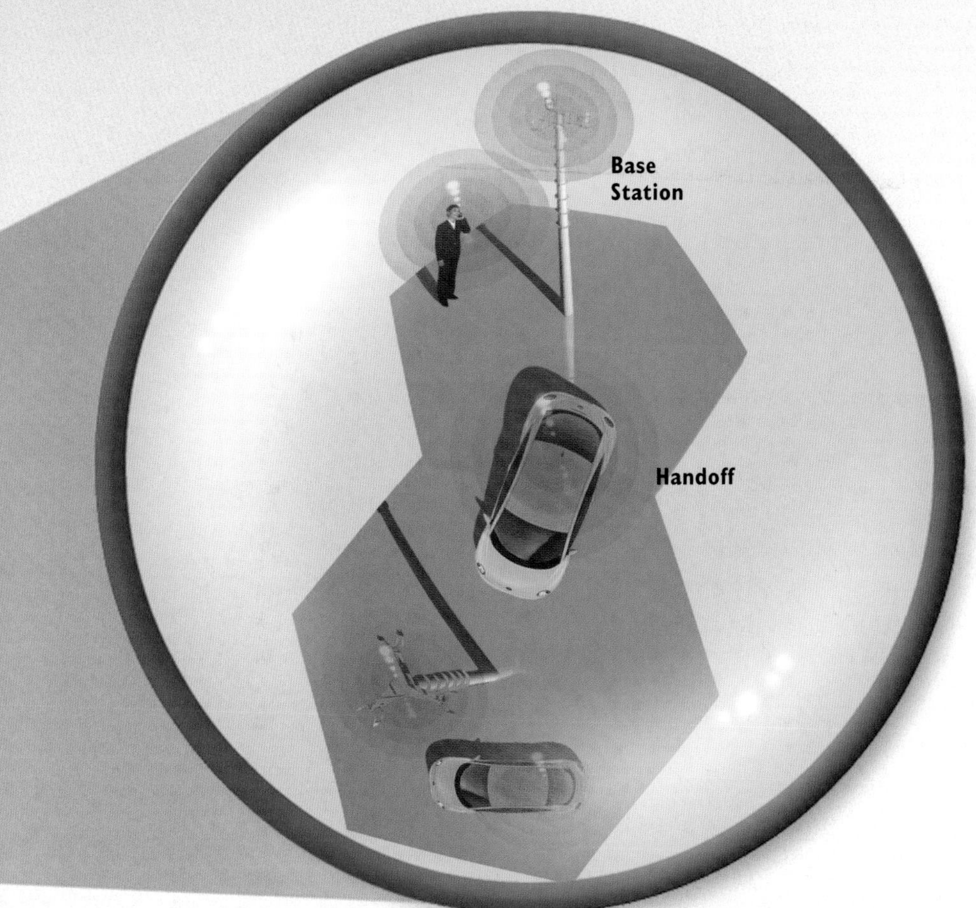

Handoff When a cell phone conversation is taking place and you move from cell to cell, a "handoff" takes place from base station to base station so that you can continue to talk.

Wireless Technologies in Our Everyday Lives

Cell Phone Here's the device that everyone thinks of when they think wireless technology—the ubiquitous cell phone.

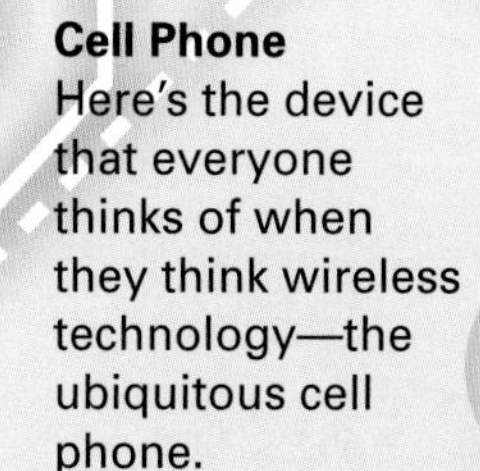

Remote-Controlled Toys Remote-controlled toys, such as cars and robots, are controlled wirelessly. And interactive toys, such as Furbies, use wireless technology to signal their presence to each other.

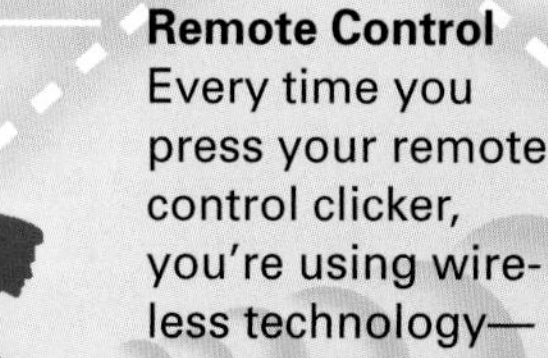

Remote Control Every time you press your remote control clicker, you're using wireless technology—infrared rays—to change the channel, change the volume, or do anything else.

Wireless Network Increasingly, homes have more than one computer—and a wireless network can enable them to communicate with one another and to share a high-speed Internet connection, such as a cable modem.

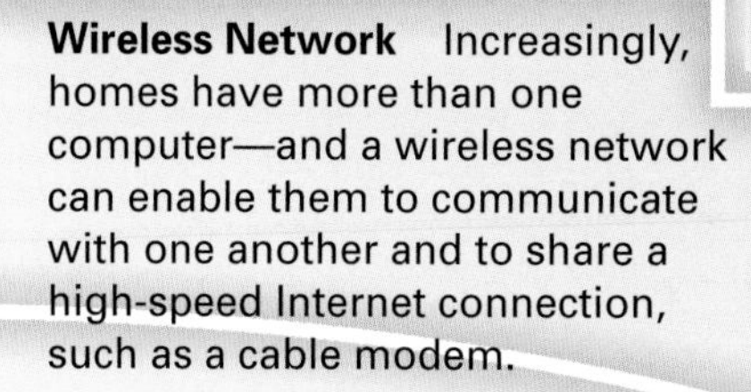

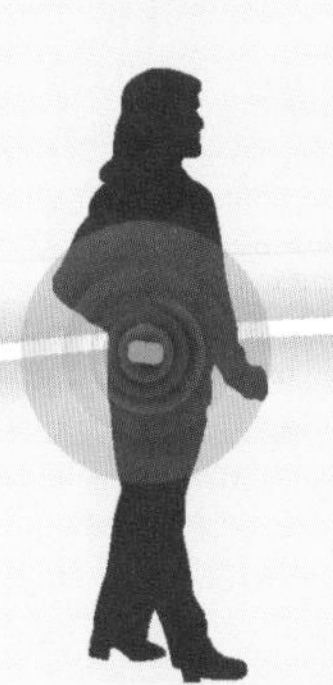

Pager When someone sends you a page, it's being sent through wireless technology.

Radio All kinds of radios, including AM radios, FM radios, and portable radios, receive signals through wireless technology.

Television TV signals are delivered wirelessly. Even if you have cable television, the television signal that you receive from your cable company was sent to the cable company wirelessly through satellites.

Palmtop Computer Palmtop computers, such as the Palm, often include cellular connections or modems so they can send and receive e-mail and other information.

Walkie-Talkies Both old-fashioned and newer walkie-talkies use wireless technology to communicate.

Wireless Tidbit

Although you don't communicate using your microwave oven, the oven is related to your cell phone and other wireless technology—it uses electromagnetic waves as a way to cook your dinner.

Garage Door Opener The next time you open your garage door, you'll be using wireless technology to open it.

CHAPTER

2

What Is the Electromagnetic Spectrum?

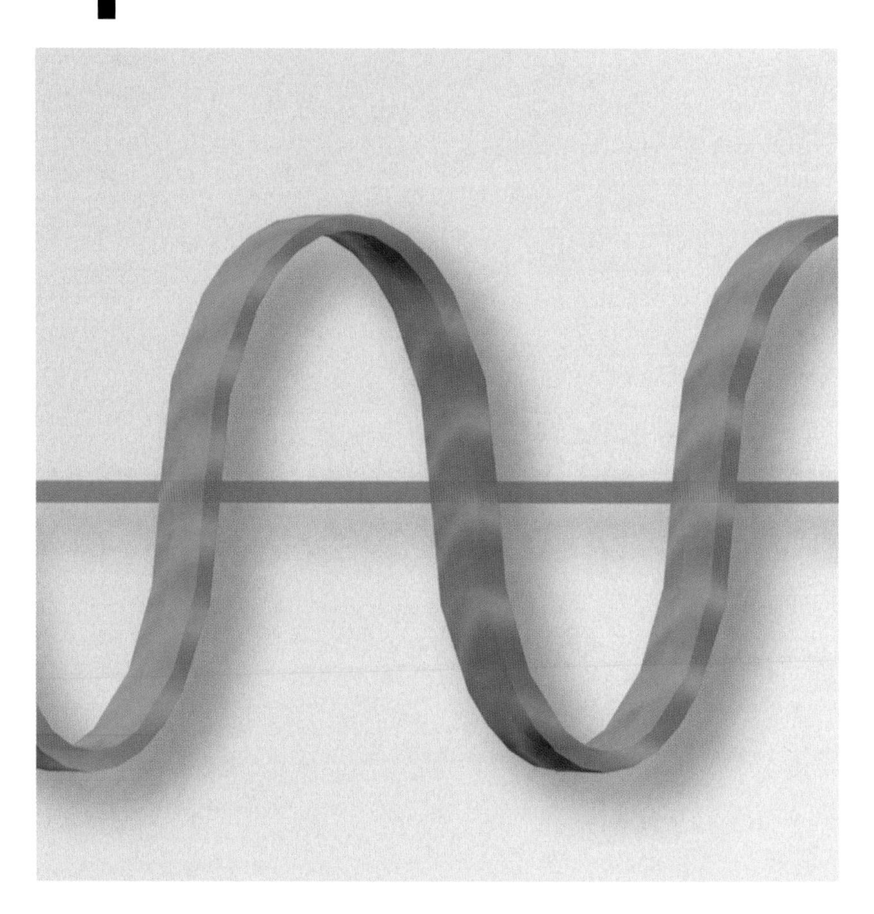

EVERY second of our lives, we are surrounded by waves of energy—some visible, the vast majority of them invisible. These waves are created in many different ways. Some, like the light and colors that we see and the different kinds of waves created by the sun, are naturally created. Others, such as radio and television signals, microwaves, remote-control infrared rays, and cell phone transmissions, are man-made.

All these waves of energy—known as *electromagnetic radiation*—taken together are referred to as the *electromagnetic spectrum*. You're probably already familiar with another kind of spectrum, the spectrum of visible light. This spectrum of visible light occupies only a very tiny portion of the electromagnetic spectrum.

To fully understand the spectrum and radiation, you must understand two basic concepts: *frequency* and *wavelength*. Wavelength, as its name implies, refers to the length of the energy wave; in other words, the length between its peaks. There are tremendous variations between wavelengths along the spectrum. They can be as long as 10^6 meters at the bottom of the spectrum, or as short as 10^{-15} meters at the top of the spectrum. (For those of you not used to the metric system, that means from distances measured in microscopic sizes, all the way up to 62 miles.)

Frequency refers to the number of times, or cycles, per second that wave cycles occur. The number of cycles per second are measured in hertz (Hz). A single cycle per second is one hertz; seven cycles per second are seven hertz. Electromagnetic waves generally go through many more cycles per second than that, though, so a shorthand is used to refer to higher numbers of hertz. One kilohertz (kHz) refers to one thousand (10^3) cycles per second; one megahertz (MHz) refers to one million (10^6) cycles per second; one gigahertz (GHz) refers to one billion (10^9) cycles per second; one terahertz (THz) refers to one trillion (10^{12}) cycles per second; and one petahertz (PHz) refers to one quadrillion (10^{15}) cycles per second. There are tremendous variations in frequency along the spectrum, with frequencies of 10^2 and below at the bottom and 10^{23} and above at the top.

As a rule of thumb, there is a relationship between frequency and wavelength: The longer the wavelength, the lower the frequency.

Radio waves are electromagnetic radiation that is capable of being used for communications. They occupy a small spot along the electromagnetic spectrum, near the bottom. They have the longest wavelengths and the lowest frequencies—characteristics that make them the most suited for sending information. The most commonly used frequencies for RF are from 9 kHz to 30 GHz.

Radio waves also can be separated into a spectrum. Because of their wavelengths and frequencies, different kinds of radio waves are suited for different kinds of communications. The higher the frequency, the shorter the range the waves can travel. The lower the frequency, the farther the range the waves can travel. So, AM radio broadcasts, for example, use a relatively low frequency, enabling them to travel long distances from their transmission towers. Cellular telephone calls use a relatively higher frequency and can travel shorter distances. They need to travel only a short distance because they need to travel only to a nearby cellular base station with which they communicate.

How Electromagnetic Waves Are Created

1 Electricity is created when electrons flow from one place to another; for example, along a wire.

2 An electric current flowing through a wire creates a magnetic field around the wire.

4 When an electromagnetic wave is generated, it takes a certain amount of time for a single cycle to complete. A single cycle is the time it takes for the current to increase and then decrease again. The frequency of the number of times a cycle completes in a second is measured in a unit called the hertz. One kilohertz (kHz) means that 1,000 cycles are completed in a second, one megahertz (MHz) means that one million cycles are completed in a second, one gigahertz (GHz) means that one billion cycles are completed in a second, and one terahertz (THz) means that one trillion cycles are completed in a second.

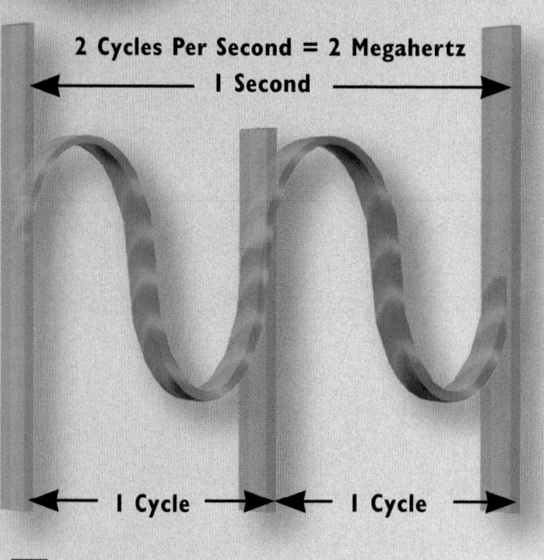

6 *Amplitude* is a measurement of the height of a wave. Amplitude is a measurement of the strength of a transmission: The higher the amplitude, the stronger the signal.

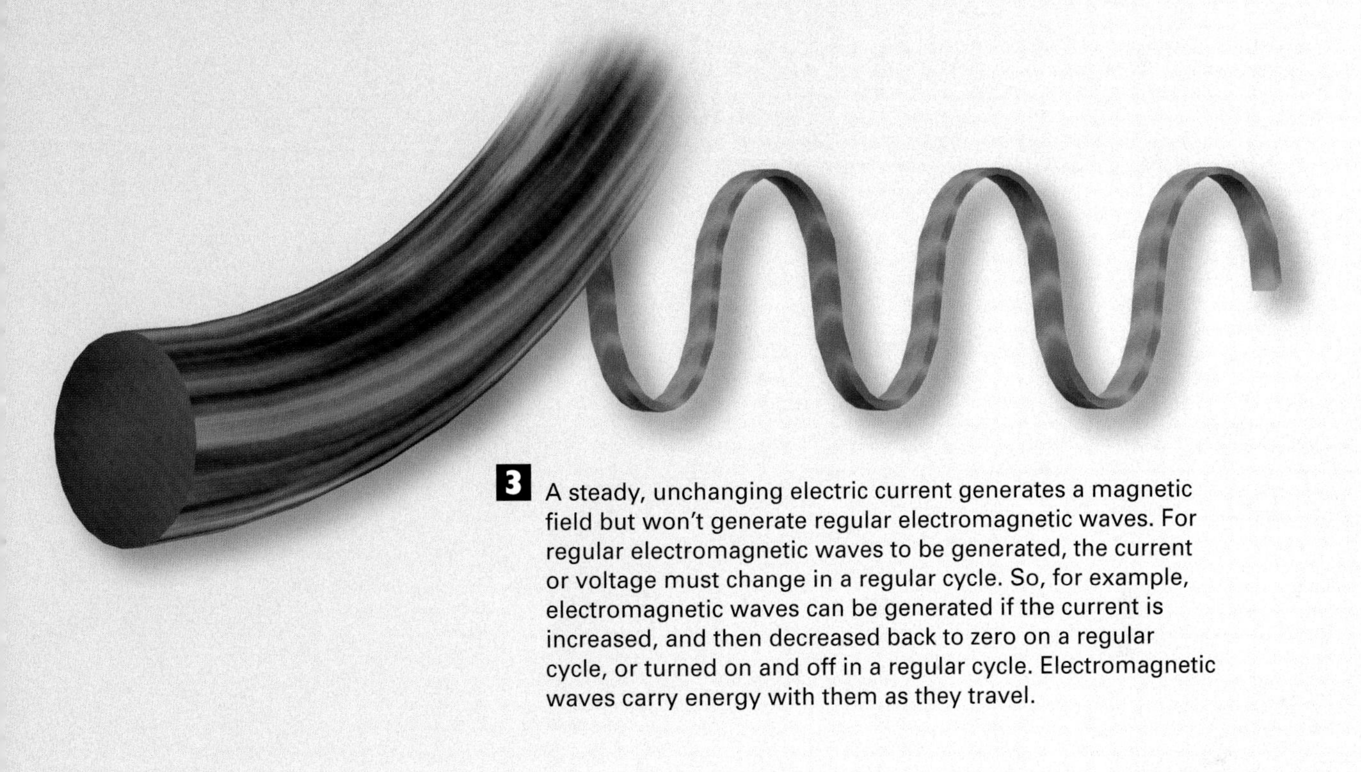

3 A steady, unchanging electric current generates a magnetic field but won't generate regular electromagnetic waves. For regular electromagnetic waves to be generated, the current or voltage must change in a regular cycle. So, for example, electromagnetic waves can be generated if the current is increased, and then decreased back to zero on a regular cycle, or turned on and off in a regular cycle. Electromagnetic waves carry energy with them as they travel.

5 *Wavelength* measures the distance between the peaks of electromagnetic waves. There is a basic relationship between wavelength and frequency: The longer the wavelength, the shorter the frequency.

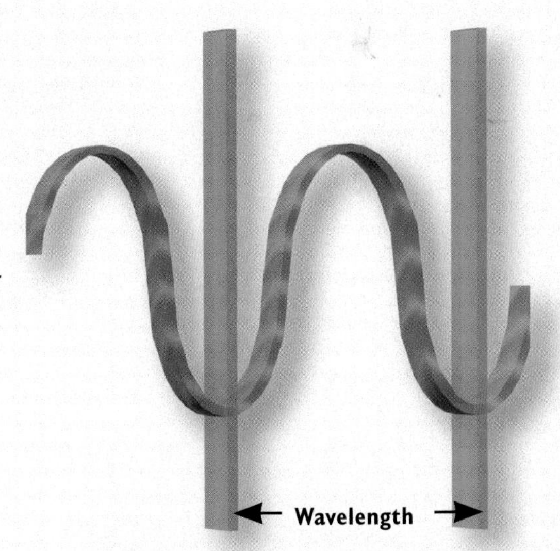

Wireless Tidbit

Electromagnetic waves travel at the speed of light—299,792,458 meters per second. Waves can be slowed down when they pass through materials, such as clouds and even the air, but the slowdown is negligible.

Understanding the Electromagnetic Spectrum

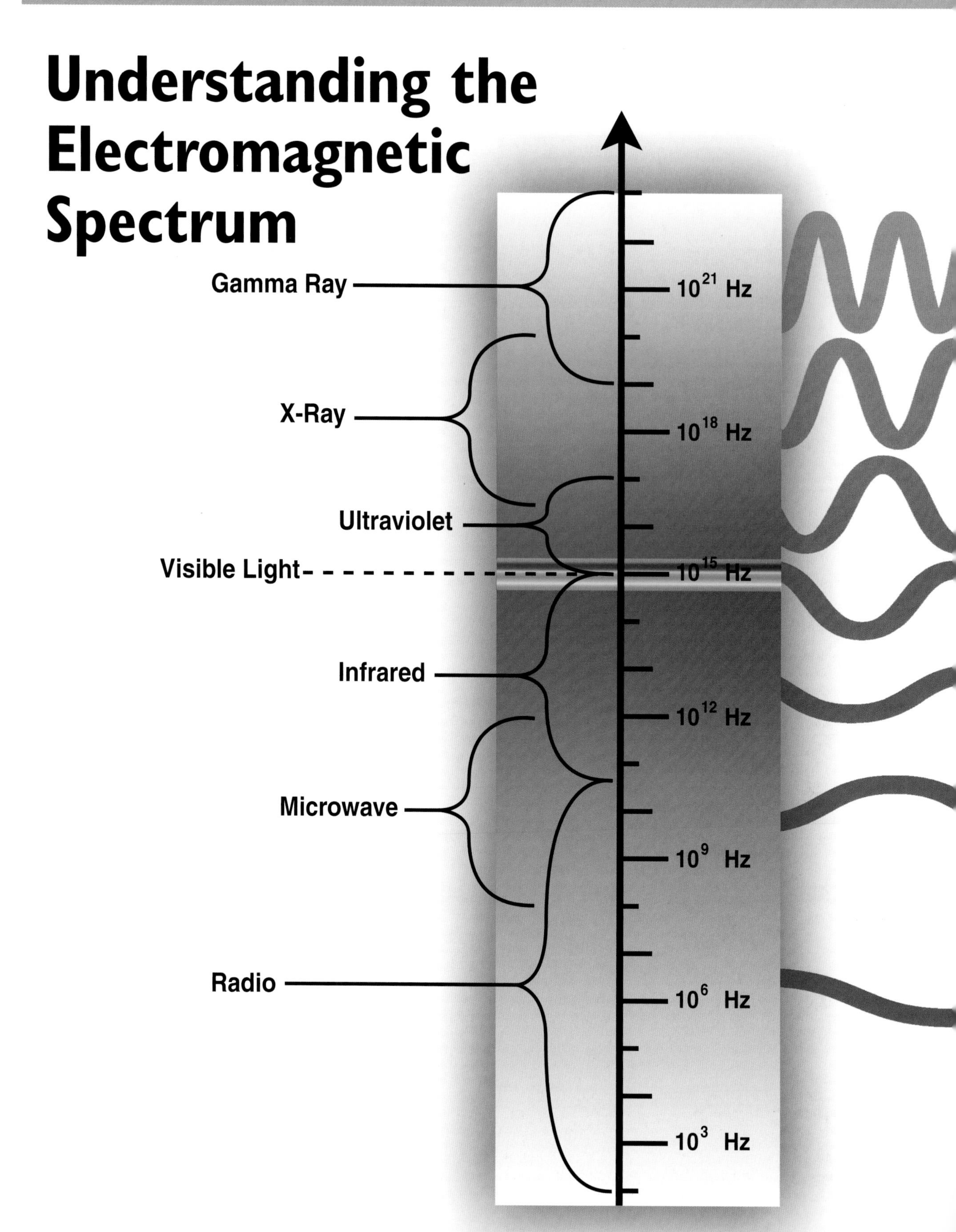

Gamma Rays All electromagnetic waves carry energy and, depending on the wavelength and frequency of the energy, the waves have different characteristics. Pictured here is the electromagnetic spectrum and what the various portions of it are used for. The highest-frequency electromagnetic waves are gamma rays, which are emitted by nuclear reactions.

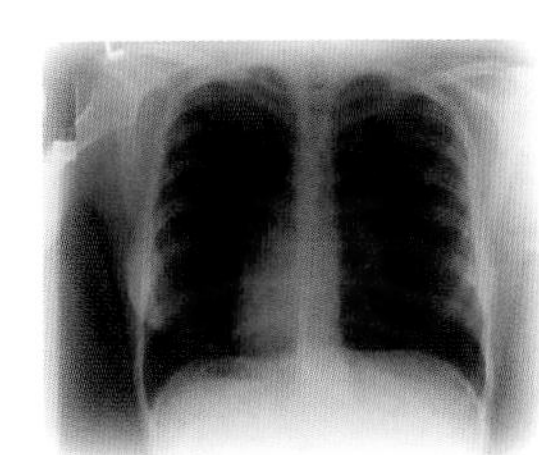

X-Rays X-rays, which can penetrate living tissue, are the next highest frequency. Their most common use is in medicine, allowing doctors to see inside the human body.

Ultraviolet Rays Ultraviolet rays are caused by, among other things, the sun. They can ionize atoms and also are dangerous to the human skin—so much so that too much exposure to them can lead to skin cancer.

Infrared Rays Infrared waves are commonly used for remote-control devices. They also can be used to let people "see" in the dark when used in night-vision devices.

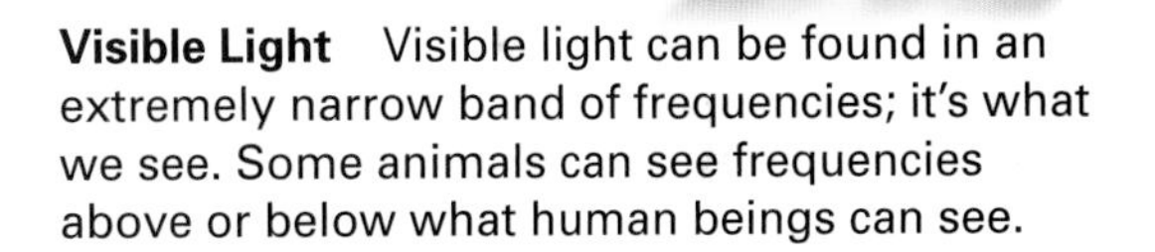

Visible Light Visible light can be found in an extremely narrow band of frequencies; it's what we see. Some animals can see frequencies above or below what human beings can see.

Microwave Microwave frequencies straddle the line between infrared and radio waves. Microwaves are used to carry communications and to cook.

Radio Frequencies Radio frequencies (RF) are at the bottom of the electromagnetic spectrum and have the lowest frequencies. *Radio* is the generic name given to electromagnetic waves that can be used for communications. Because new technologies are continually being developed, the range of frequencies that can be used for communications is constantly expanding.

Wireless Tidbit

Cycles per second are called hertz (Hz). The term is named after Heinrich Hertz, a German physicist who discovered radio waves and was the first person to broadcast and receive those waves.

Understanding the Radio Frequency Spectrum

There are several bands of frequencies within the RF spectrum and, because of the physical characteristics of waves within those bands, they're used for different kinds of communications. Higher frequencies are more easily blocked by physical objects, whereas lower frequencies can penetrate them. Higher frequencies, however, also carry more energy. So, for example, visible light is blocked by walls and houses, but lower-frequency RF waves can penetrate through them, which is why RF waves are used for communications.

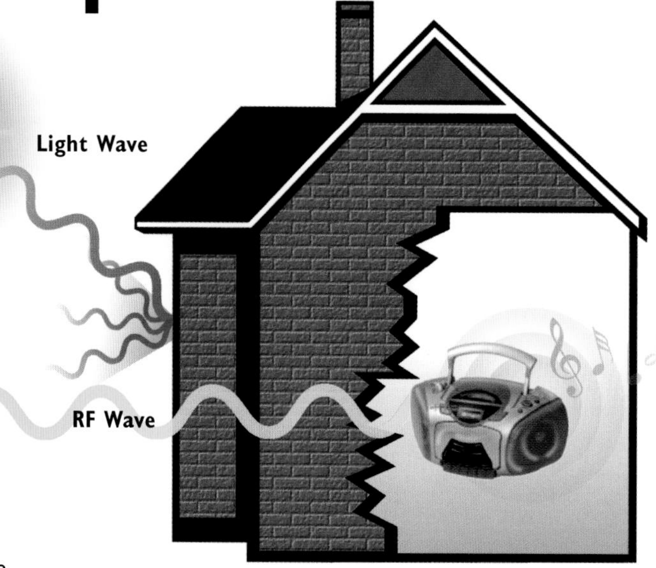

Extremely Low Frequency Extremely Low Frequency (ELF) waves below 3 kHz are used for submarine communications.

Very Low Frequency Very Low Frequency (VLF) waves between 3kHz and 30 kHz are used in maritime communications.

Low Frequency and Medium Frequency Low Frequency (LF) or Long Wave (LW) between 20 and 300 kHz are used in AM radio broadcasting. Medium Frequency (MF) or Medium Wave (MW) waves between 300 and 3,000 kHz are used in AM radio broadcasting as well.

High Frequency High Frequency (HF) or Short Wave (SW) waves between 3 and 30 MHz are used in AM broadcasting and in shortwave and amateur radio.

Very High Frequency Very High Frequency (VHF) waves between 30 and 300 MHz are used in FM radio and television broadcasting.

Ultra High Frequency Ultra High Frequency (UHF) waves between 300 and 3,000 MHz are used in television broadcasting and by cellular telephones.

Super High Frequency Super High Frequency (SHF) waves between 3 and 30 GHz are used in fixed wireless communications and for satellite transmissions.

Extremely High Frequency Extremely High Frequency (EHF) waves between 30 and 300 GHz are used for satellite transmissions and for radar.

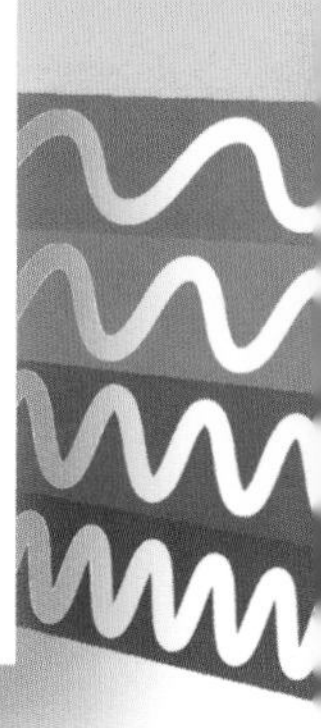

2 Higher-frequency waves travel shorter distances than do lower-frequency waves. That's why lower-frequency waves are used for radio broadcasting, for example. Higher-frequency waves are used for cellular telephone networks. Because these waves don't travel great distances, you can use the same bandwidth for different calls in areas relatively close to each other, and the calls won't interfere with one another.

CHAPTER

3

How a Basic Wireless System Works

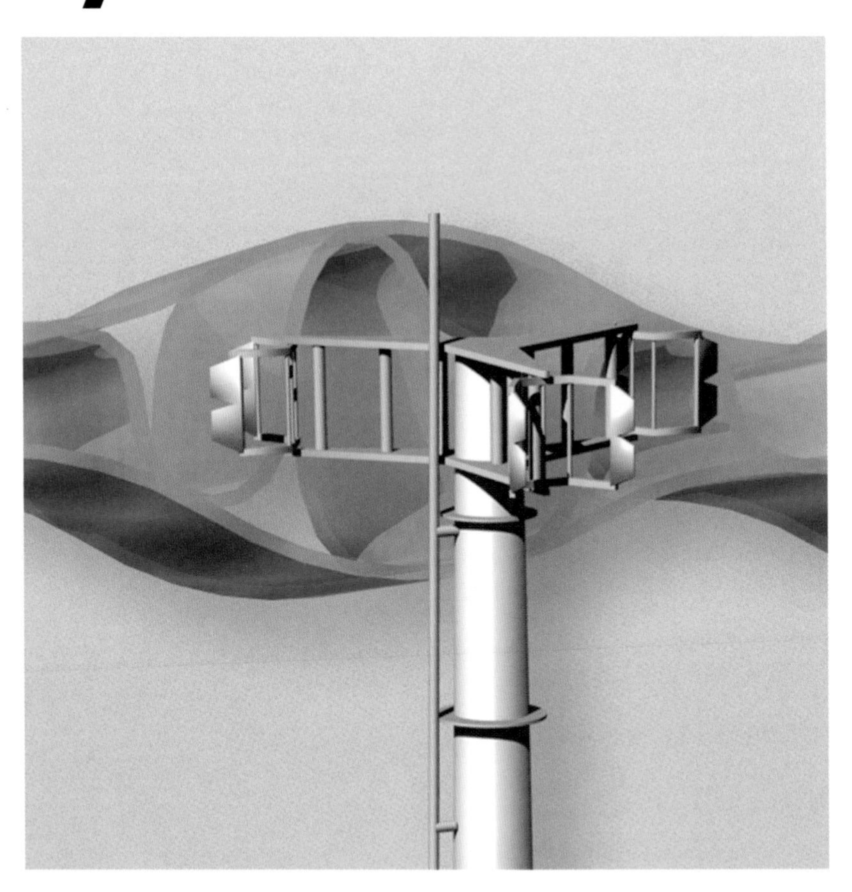

AS you learned in Chapter 2, radio frequency (RF) waves—waves that make up a small part of the electromagnetic spectrum—are used to send wireless information from one device to another, such as cellular telephones or a television. But how does a basic wireless system work? How do computer data, television transmissions, or the spoken voice over the telephone get from point A to point B without the use of wires?

No matter how simple or complex the system, and no matter what kind of information is being transmitted, the basic wireless system for transmitting information remains very much the same. At its heart, it's fairly simple—it's the details that are complicated. First, the information that is to be transmitted needs to be created. Next, it's encoded onto a radio wave, and then it's transmitted.

The signal, now in wave form, travels through the air and is ultimately received by an antenna or aerial, which sends it along to a receiver. Finally, a variety of devices transform the energy in the signal into the electrical energy that can be recognized by the receiving device, whether it be a television set, handheld computer, or cellular telephone.

It all sounds so simple, but as you'll see throughout this book, there are complications and endless variations on this one theme. Networks can bounce RF waves off satellites and deliver them tens of thousands of miles, or can transmit radio broadcasts into a single neighborhood. They can be used to create vast wireless systems connecting thousands of computers in major corporations, or they can connect two computers in your home office. They can send TV, radio, or voice signals; they can help us search for life other places in the universe. They can carry voice or data of all different kinds.

The use of the word "network" with regard to wireless technology is a loose one. Generally, though, it means a system in which RF signals carrying information of some kind are sent from one device, through intervening communications devices, and then to a receiving device. So, for example, a simple child's walkie-talkie wouldn't be considered a network, because with walkie-talkies, the signal is sent directly from one device to another. But cell phones are part of a network, because they don't communicate directly with other telephones—they are hooked into a huge network that routes calls using many different pieces of hardware and software.

As you'll see throughout this book, the most important wireless technologies in one way or another use networks. These networks have become increasingly complicated over time. But keep in mind, as you learn about them, that at heart they're all the same simple system: Information gets created, encoded onto RF waves, sent through intervening communications devices, and then received, decoded, and used.

How a Basic Wireless Network Works

1. Many kinds of information can be transmitted wirelessly, including computer data, voice phone calls, TV and radio transmissions, and more. So, first, the information to be transmitted comes from a device, such as a handheld computer, radio station, or cell phone.

2. For the information to be sent, it must be piggybacked onto an RF wave (also called a *signal*) in a process called *modulation*. The signal on which the information will be sent is called a *carrier wave*. The information is put onto the carrier wave by a *modulator*, a device that can use a variety of methods to superimpose the information onto the carrier wave. Note that a modulator might be built into the device that creates the data, such as the case with a cell phone or a handheld computer. However, it might be separate from the device that creates the data, such as in TV broadcasts. (For more information about modulation, see Chapter 4, "How Amplitude Modulation (AM) and Frequency Modulation (FM) Work.")

3. The signal now needs to be sent. It's sent through a transmitter that takes the signal and sends it through the air. Depending on what needs to be transmitted, the distance it needs to be transmitted, and the strength that the signal has to be, the transmitter can be a variety of different sizes. It can be small, like the built-in antenna on a cell phone, or large, like a television transmitter on top of a high tower.

Data

Modulator

Transmitter

7. The information is sent to the receiving device, such as a cell phone, TV set, or handheld computer, which can now display the information.

6. A modulator (also called a demodulator) interprets the signal and separates the carrier wave from the information that was sent on the wave. It translates the information back into its original form.

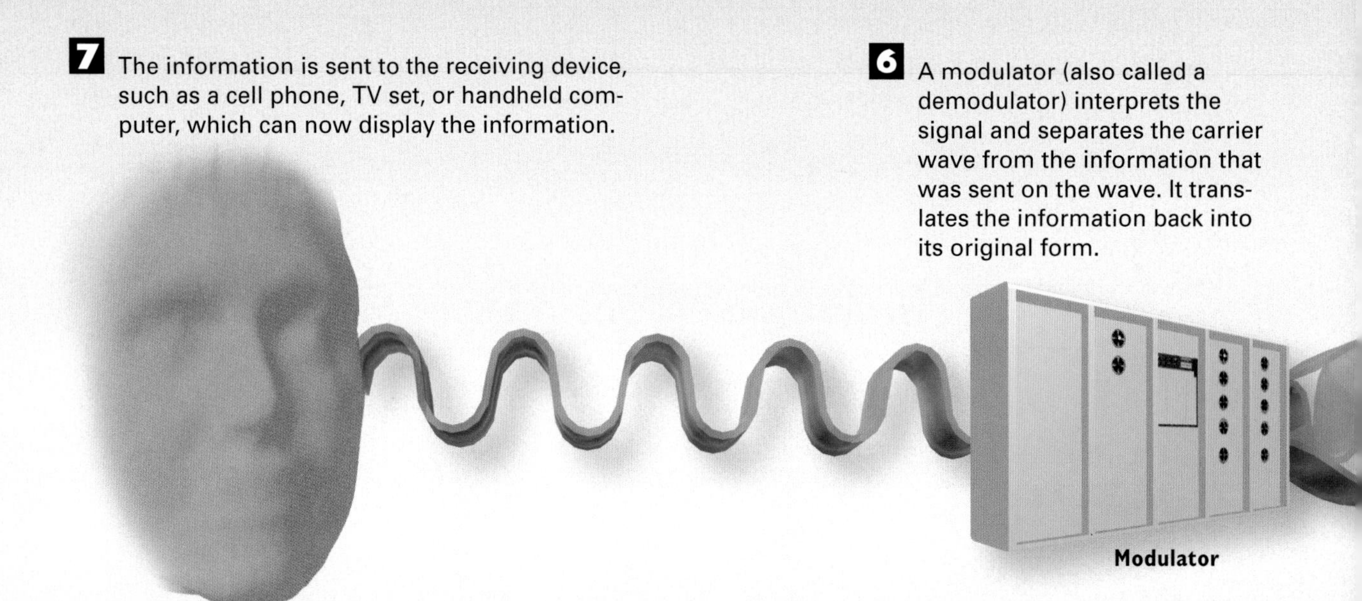

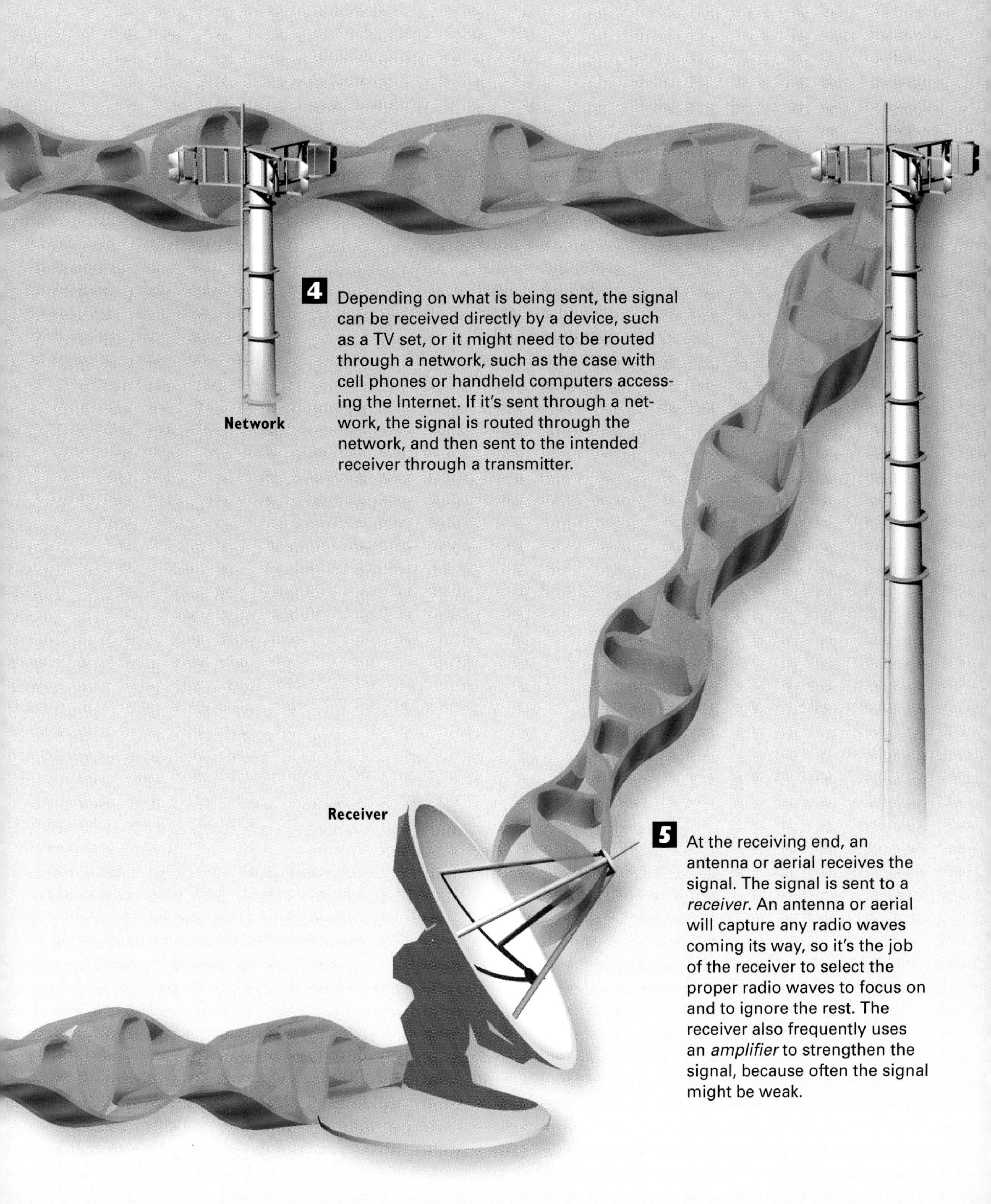

4 Depending on what is being sent, the signal can be received directly by a device, such as a TV set, or it might need to be routed through a network, such as the case with cell phones or handheld computers accessing the Internet. If it's sent through a network, the signal is routed through the network, and then sent to the intended receiver through a transmitter.

5 At the receiving end, an antenna or aerial receives the signal. The signal is sent to a *receiver*. An antenna or aerial will capture any radio waves coming its way, so it's the job of the receiver to select the proper radio waves to focus on and to ignore the rest. The receiver also frequently uses an *amplifier* to strengthen the signal, because often the signal might be weak.

CHAPTER 4

How Amplitude Modulation (AM) and Frequency Modulation (FM) Work

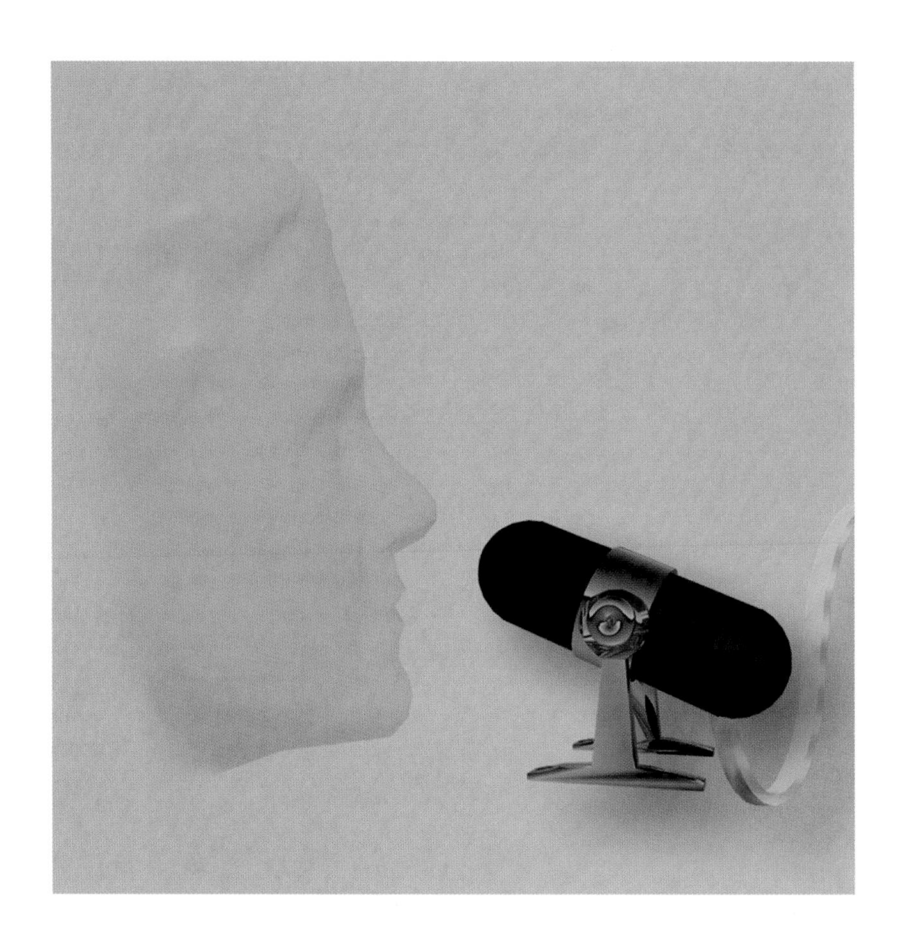

INFORMATION that is going to be sent wirelessly needs some way to get from one place to another through the air. To get from point A to point B, it rides on the back of RF waves. The wave that carries the information is called a *carrier wave*.

Information is put onto the carrier wave through a process called *modulation*. Modulation takes the information, which can be in analog or digital form, and superimposes it onto a carrier wave. (Digital data is data represented as either on or off, like the data in a computer. Analog data is information represented along a continuum—there can be infinite variations between two points.) The carrier wave itself is always an analog wave. The carrier wave then is sent through the air, and on the other end, it is *demodulated*—that is, the information is separated from the carrier wave.

The two primary means of modulation are *amplitude modulation (AM)* and *frequency modulation (FM)*. You're no doubt familiar with both terms from your radio and, as you've probably guessed by now, AM broadcast radio uses amplitude modulation, and FM broadcast radio uses frequency modulation. These types of modulation aren't limited to radio broadcasts, however—one way or another, all wireless communications uses some form of AM or FM modulation.

In AM, the frequency of the carrier wave stays constant, but the *amplitude* of it—in other words, its height—changes as a way to represent the information being sent. AM is the earliest kind of modulation, and it has been around since the earliest days of radio communications. It's easy to implement, but there's a problem with it—it's prone to interference, and creating a high-quality signal using AM is difficult. There are several reasons for that, primarily that it's easier for interference to change the amplitude of a signal than it is for it to change the frequency of a signal. That means that AM suffers more from interference than FM.

A new generation of AM, called *Binary Amplitude Shift Keying (BASK)*, is more resistant to interference and noise, and it can be used to transmit digital data. Because of that, it's used in some digital wireless systems.

FM works differently than AM. In it, the amplitude of the carrier wave stays constant, but the frequency changes—in other words, the speed at which it goes through a wave cycle constantly changes. The changing frequency is what represents the information being sent. Noise and interference don't affect the frequency of RF signals in a major way, so because of that, the quality of the FM signal tends to be higher than the quality of AM signals.

Many kinds of digital wireless communications use a variant of FM, called *Phase Modulation (PM)*. PM uses the fact that there are different points in a wave cycle to transmit information. It continually shifts the points in the cycle—and each of those points can represent different information. It's particularly well-suited for representing digital information. Variants of it are used in many major cellular technologies.

How AM Works

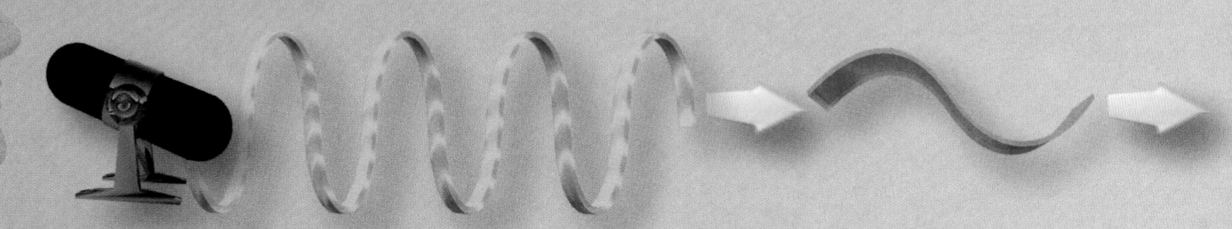

1 AM works by varying the amplitude (height) of the carrier signal as a way to carry the information being transmitted, but the frequency stays the same. First, information, such as voice communications, is created as an analog wave.

2 A *modulator* superimposes the information wave onto a carrier wave. The modulator works by using an oscillator to create a wave, which the modulator combines with the information wave, resulting in the carrier wave.

5 A variant of AM called *Binary Amplitude Shift Keying (BASK)* can transmit digital data and is less sensitive to noise, so it's used in some digital wireless systems. In it, digital data is superimposed onto a carrier wave, and the resulting signal resembles the original digital data. The signal is more resistant to noise because the receiving equipment needs to differentiate between only two states—on and off—and does not need to interpret widely varying amplitudes. Because BASK uses digital data, it requires special digital processing chips, which is one of the reasons why it has come into use so much more recently than AM or FM.

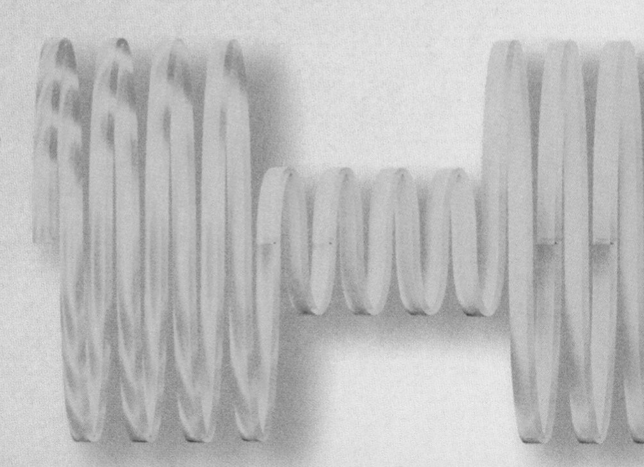

3 In the resulting signal, the varying amplitude represents the information being transmitted.

4 One problem with using AM as a way of transmitting information is that it's prone to interference and noise, because interference and noise can cause random changes in the amplitudes of waves. Because of this, AM transmissions—such as AM radio—tend to be of low quality.

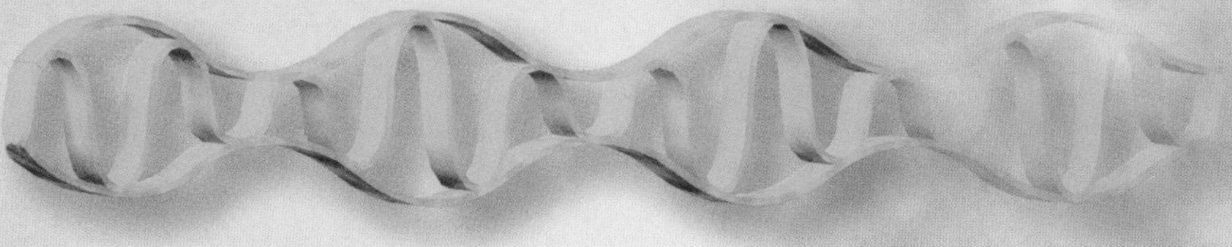

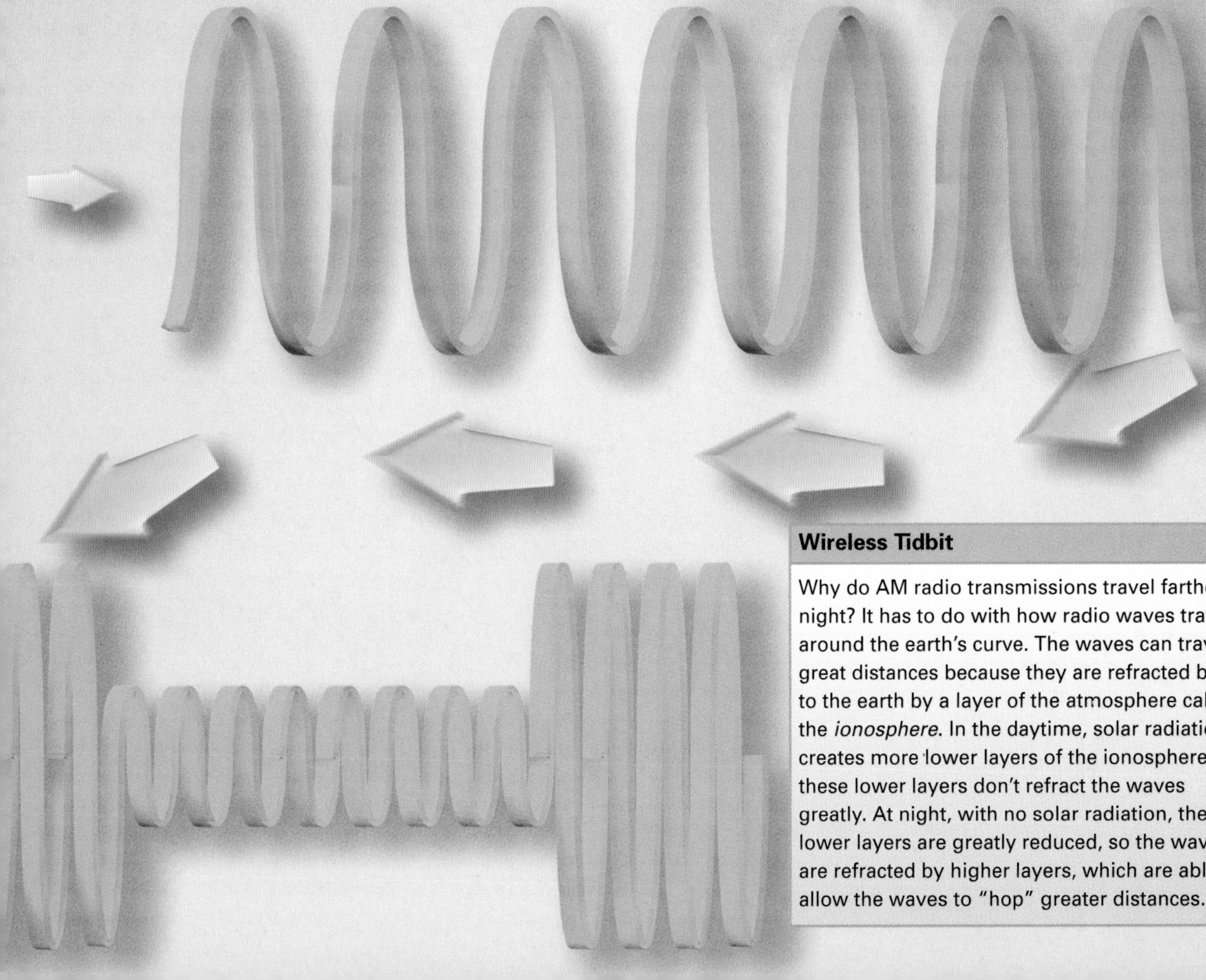

Wireless Tidbit

Why do AM radio transmissions travel farther at night? It has to do with how radio waves travel around the earth's curve. The waves can travel great distances because they are refracted back to the earth by a layer of the atmosphere called the *ionosphere*. In the daytime, solar radiation creates more lower layers of the ionosphere, and these lower layers don't refract the waves greatly. At night, with no solar radiation, these lower layers are greatly reduced, so the waves are refracted by higher layers, which are able to allow the waves to "hop" greater distances.

How FM Works

1. FM works by varying the frequency of the carrier signal as a way to carry the information being transmitted, but the amplitude (height) stays the same. In the first step, information, such as voice communications, is created.

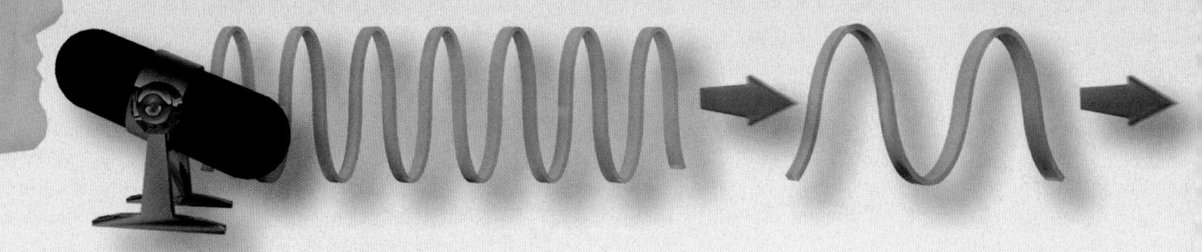

4. A variant of FM, *Phase Modulation (PM)*, is used in many kinds of digital wireless communications. It doesn't use the entire wave signal as a way of sending information. Instead, it sends information by suddenly shifting which phase of the wave is sent at a particular time. To understand how it works, you first must know the four phases of a wave cycle, represented as 0°, 90°, 270°, and 360°.

2 The information wave is superimposed upon a carrier wave. In the resulting signal, the varying frequency represents the information being transmitted. The modulator works by using an oscillator to create a wave that the modulator combines with the information wave, resulting in the carrier wave.

3 FM is less susceptible to noise and interference than AM because it doesn't matter whether the amplitude of the signal changes—the receiver only pays attention to the frequency, not the amplitude. Because of that, the FM signal can use extra bandwidth to get more information into the signal. As a result, higher-quality information can be transmitted—it can transmit music in stereo, for example, whereas AM transmits only in mono because of interference problems.

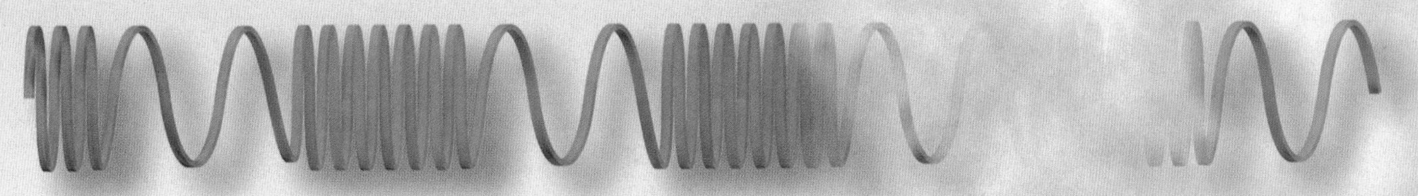

5 PM uses *phase-shift keying* as a way of conveying information. In it, the information is represented by the phase of the wave at a particular point in time. Therefore, if the phase begins at 270°, for example, it means one piece of information, whereas if the phase begins at 90°, it means another piece of information. Many of the newest digital cellular technologies use variants of this technique. This kind of modulation has been used only in the past 10 years or so because it requires sophisticated electronics components in its modulators.

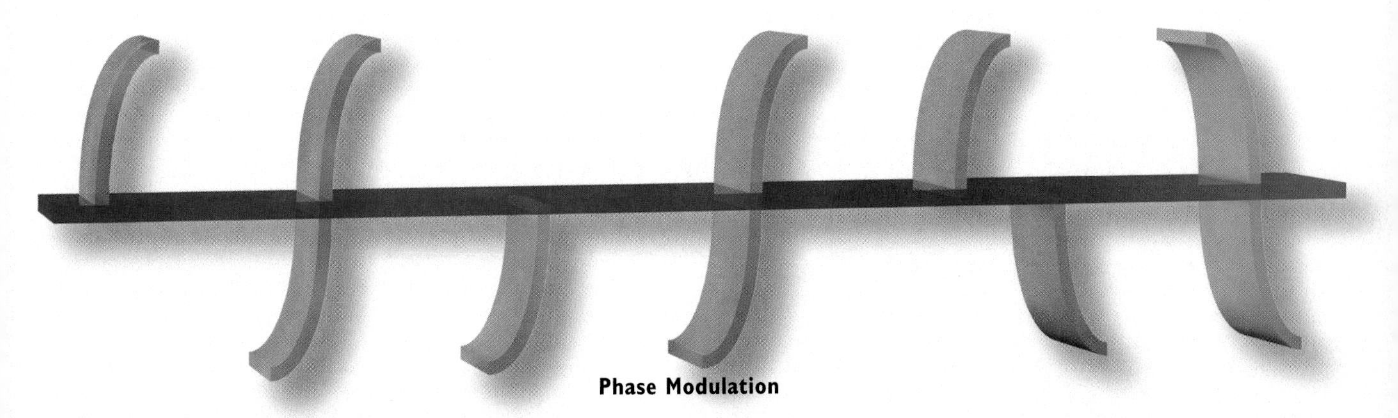

Phase Modulation

CHAPTER

5

How Data Rides on Wireless Waves

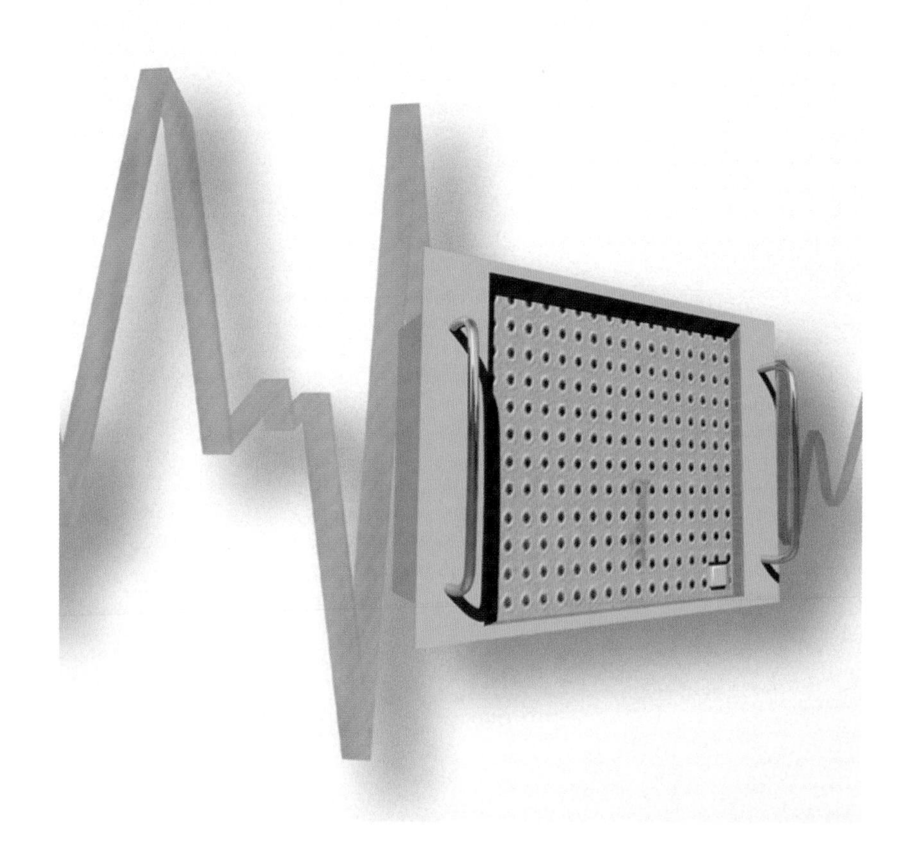

YOU know by now that wireless communication requires information to be sent along RF waves, but how is that information actually transmitted along waves? That's what you'll find out in this chapter.

As you learned in previous chapters, for information to be transmitted wirelessly, it first needs to be modulated onto a carrier wave. The information to be transmitted can be of many different kinds—radio, television, voice, or data, for example. But no matter what kind it is, it can be of two different types, either *analog* or *digital.*

Analog data is information represented along a continuum—there can be infinite variations between two points. So, for example, a watch face that has hands on it represents data in an analog manner. And a wave itself is analog, because it's continuous.

Digital data, on the other hand, is information presented as either on or off, often represented as 1 for on and 0 for off. All data in computers is digital data.

Whether the data is digital or analog, when it's transmitted wirelessly, it rides on RF waves, which are analog. So, even digital data has to piggyback onto an analog wave to be transmitted.

Until recently, all wireless data, such as TV and radio transmissions and conversations sent via cell phones, was analog. But increasingly, data is sent in digital format. Digital data sent wirelessly is superior to that sent in analog format for a variety of reasons. It can be more efficiently sent, it's easier to be sure that the data hasn't been corrupted during the transmission, and it's easier to encrypt it so that eavesdroppers can't listen in, among other reasons. Because of this, the newest wireless technology is digital, including new, high-speed cellular telephone services, wireless Internet access, and digital television.

No matter what kind of data is to be transmitted, it frequently needs to be processed in some way before it's sent. That's because of the very nature of RF waves and the environment through which they travel. The environment is full of electromagnetic radiation caused by many different things, such as the normal background radiation caused by the sun, sunspots, machinery, and lightning storms. This electromagnetic radiation is called *noise*, because it doesn't carry information. Weak RF signals would be drowned out by this noise if they weren't in some way strengthened before they were sent.

Another problem with sending information wirelessly is that the environment naturally weakens RF waves as they travel. Everything with which the RF waves come in contact, such as air molecules, rain, buildings, even leaves on trees, weakens the waves through a process called *absorption.* The waves also can be deflected by objects they come into contact with. Because of this, RF waves often need to be processed before they're sent—changed so that they can be transmitted efficiently, and strengthened so that they can reach their destination. Devices such as signal processors and amplifiers process and strengthen the signal.

How Data Is Transmitted by Waves

1. The information to be transmitted is put onto a carrier wave through *modulation*. (See Chapter 4, "How Amplitude Modulation [AM] and Frequency Modulation [FM] Work," for information on how modulation works.)

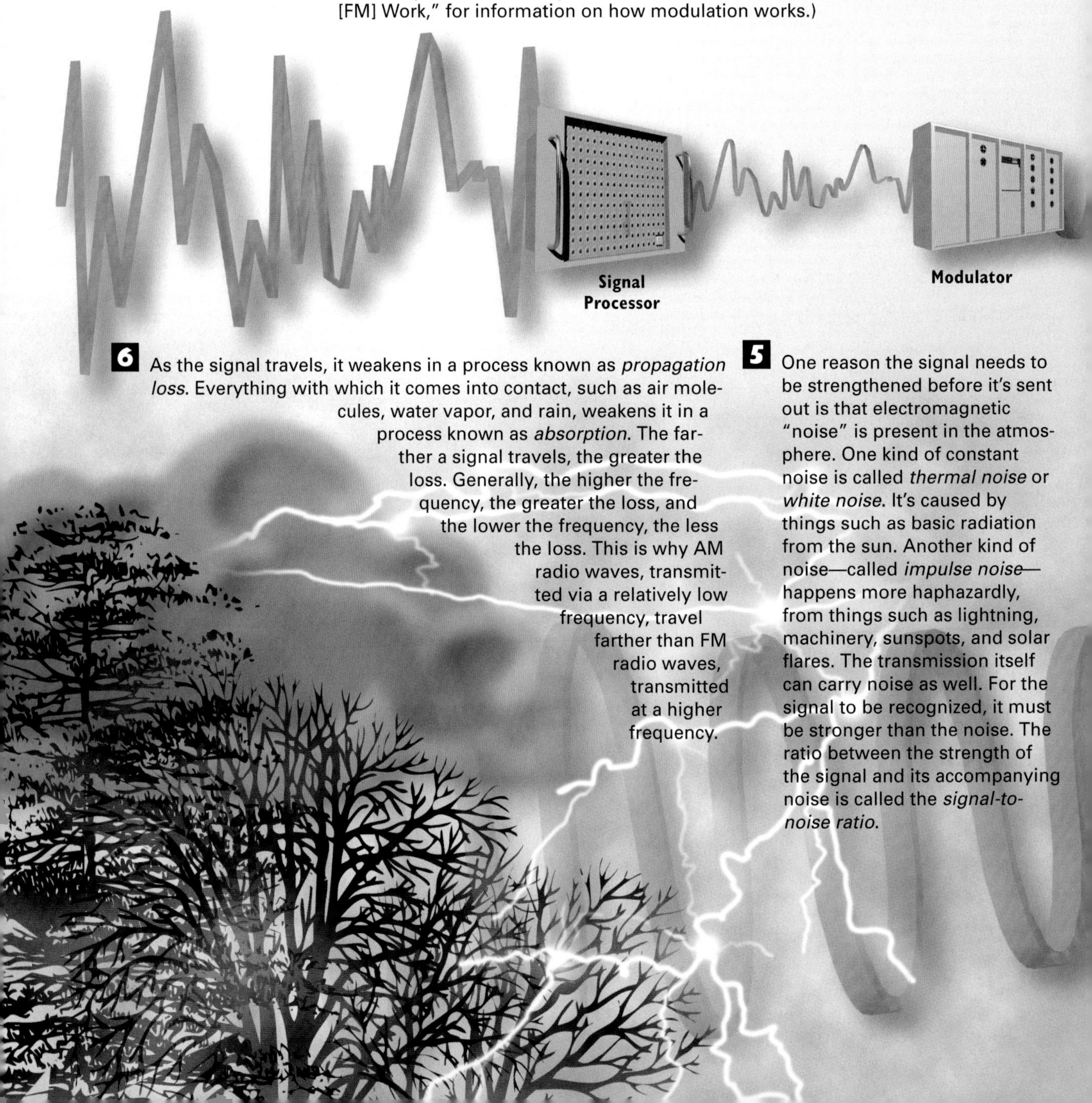

Signal Processor

Modulator

5. One reason the signal needs to be strengthened before it's sent out is that electromagnetic "noise" is present in the atmosphere. One kind of constant noise is called *thermal noise* or *white noise*. It's caused by things such as basic radiation from the sun. Another kind of noise—called *impulse noise*—happens more haphazardly, from things such as lightning, machinery, sunspots, and solar flares. The transmission itself can carry noise as well. For the signal to be recognized, it must be stronger than the noise. The ratio between the strength of the signal and its accompanying noise is called the *signal-to-noise ratio*.

6. As the signal travels, it weakens in a process known as *propagation loss*. Everything with which it comes into contact, such as air molecules, water vapor, and rain, weakens it in a process known as *absorption*. The farther a signal travels, the greater the loss. Generally, the higher the frequency, the greater the loss, and the lower the frequency, the less the loss. This is why AM radio waves, transmitted via a relatively low frequency, travel farther than FM radio waves, transmitted at a higher frequency.

2 Depending on the kind of information being transmitted, the signal might need to undergo *signal processing* so that the signal can be transmitted more effectively. In the case of an audio transmission, for example, many frequencies in the signal can be eliminated because the human ear can't hear high and low frequencies—so a signal processor eliminates them. Audio signal processors process audio transmissions, and digital signal processors process digital transmissions. There are many different kinds of signal processors, and they use many different types of technologies to do their work—notably, computer chips.

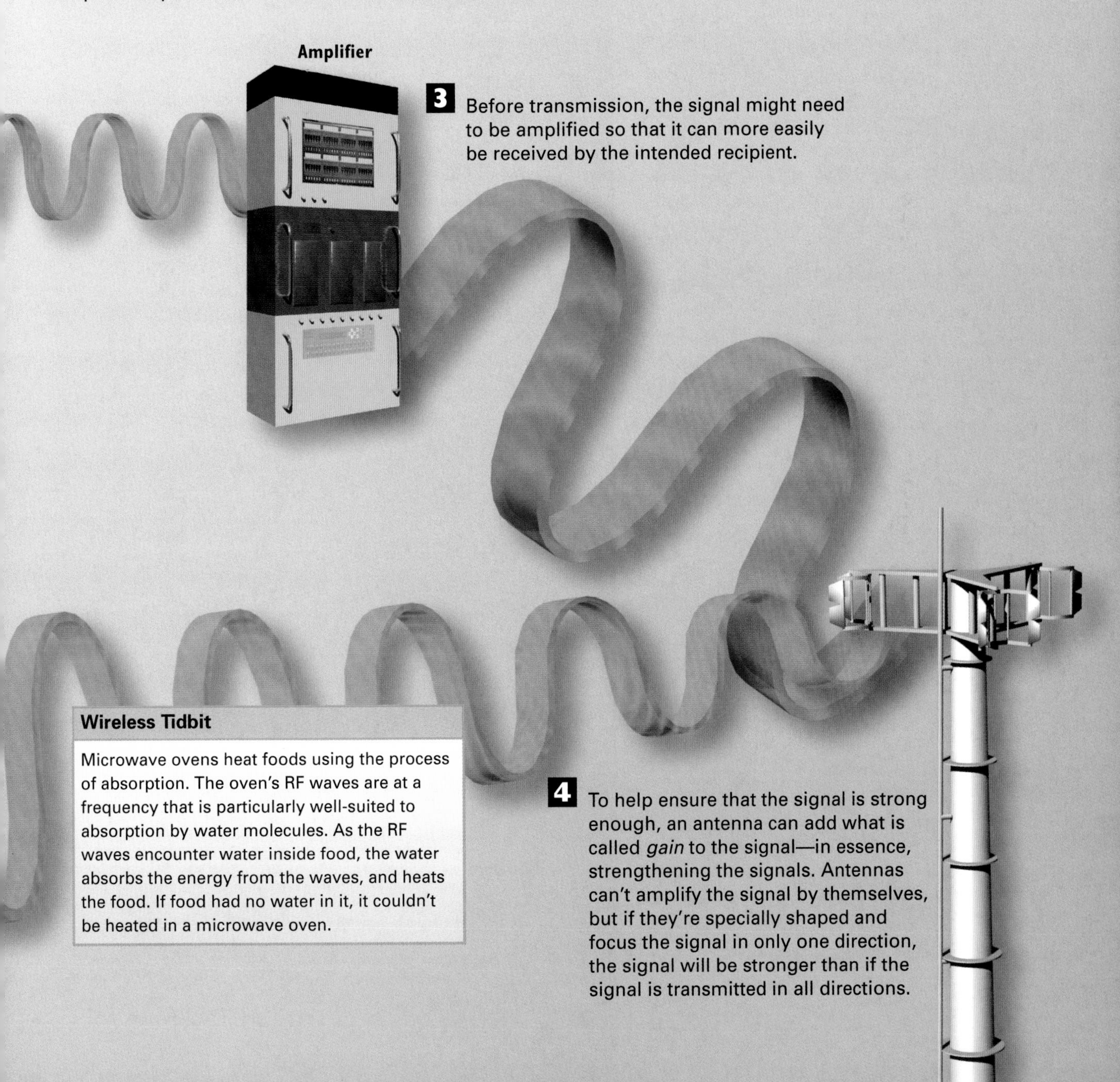

3 Before transmission, the signal might need to be amplified so that it can more easily be received by the intended recipient.

Wireless Tidbit

Microwave ovens heat foods using the process of absorption. The oven's RF waves are at a frequency that is particularly well-suited to absorption by water molecules. As the RF waves encounter water inside food, the water absorbs the energy from the waves, and heats the food. If food had no water in it, it couldn't be heated in a microwave oven.

4 To help ensure that the signal is strong enough, an antenna can add what is called *gain* to the signal—in essence, strengthening the signals. Antennas can't amplify the signal by themselves, but if they're specially shaped and focus the signal in only one direction, the signal will be stronger than if the signal is transmitted in all directions.

Understanding Digital and Analog Signals

1 In an *analog* signal, the wave's amplitude changes continually over time. The signal can have an infinite number of amplitudes among any two points. Think of a light dimmer as an analog signal—the light intensity can vary endlessly.

3 Data transmissions have many benefits over analog transmissions, including greater reliability, noise reduction, greater security, and the capability to carry more kinds of services in a single transmission. Although some data, such as computer information, is created digitally, other kinds of data, such as the human voice, must be converted from analog to digital.

4 In the first step of conversion, after someone speaks, a microphone or other kind of device converts the voice into analog electrical waves.

5 The waves are sent through a digital signal processor, which takes out those parts of the wave that are beyond or below the range of human hearing.

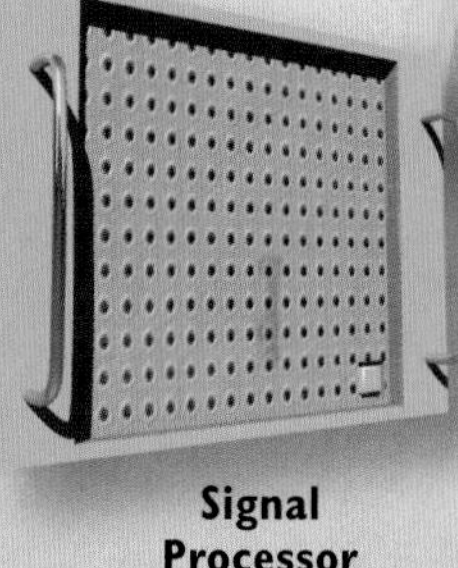

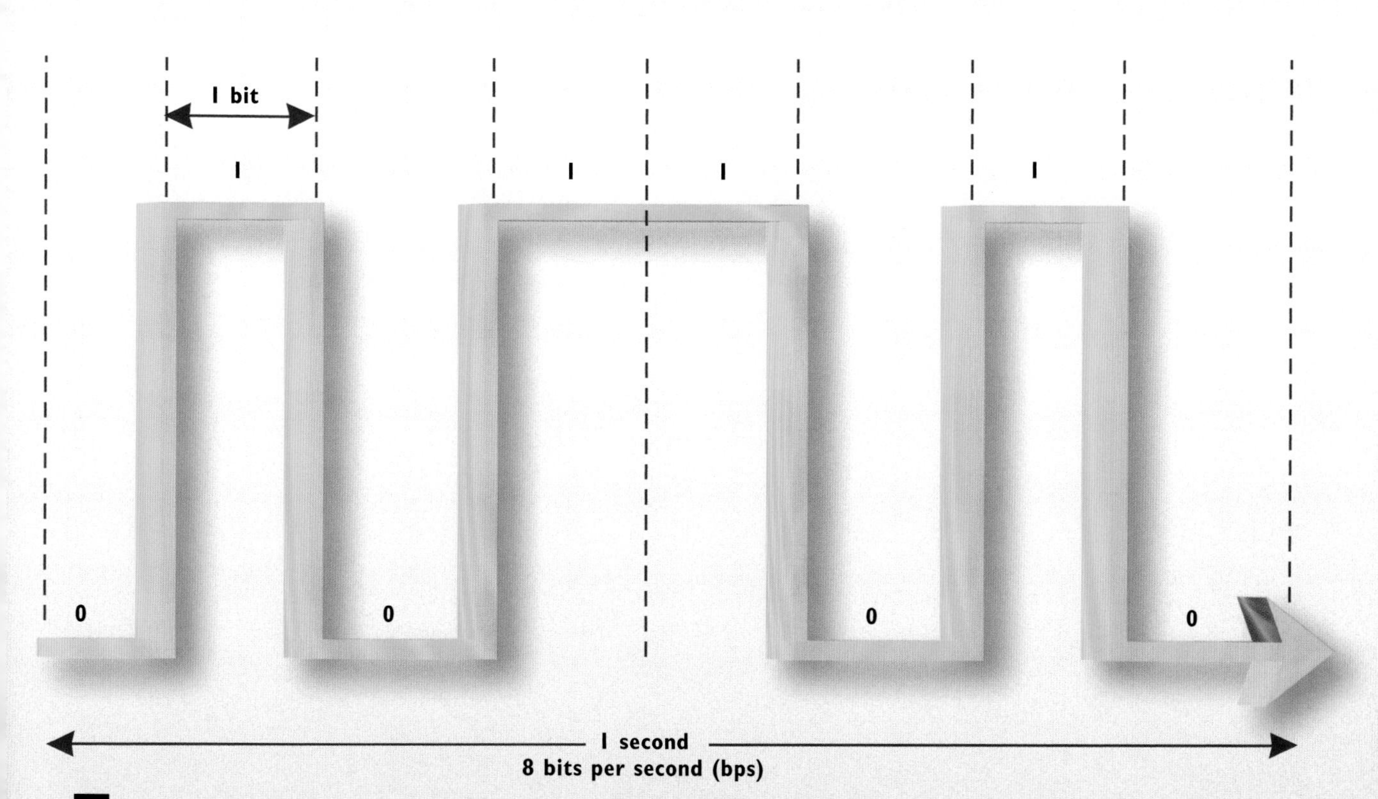

2 In a *digital* signal, there is no continuum—information is presented as either on or off, often represented as 1 for on and 0 for off. A single piece of data—again, either on or off—is called a *bit*. Eight bits make up a single byte. Computers process data digitally. The speed of transmission of digital data generally is represented as bits per second (bps).

6 Next, the processed wave is sent through an analog-to-digital converter, which samples the wave at a certain number of times per second and converts the analog wave into digital information. The more times per second that the save is sampled, the greater the quality of the digital information. The resulting digital information now can be transmitted—but when it does, it will be sent on a carrier wave, which is an analog wave. On the receiving end, the device will be able to separate the digital data from the carrier wave.

Wireless Tidbit

One of the earliest methods of long-distance communication used an early form of digital communication: Morse code. Morse code represents the letters of the alphabet by using an arrangement of dots and dashes—in essence, the 1s and 0s of digital communications. So, when telegraph operators tapped out Morse code, they were the forerunners of the digital revolution.

Audio-to-Digital Converter

CHAPTER

6

How Antennas, Transmitters, and Receivers Work

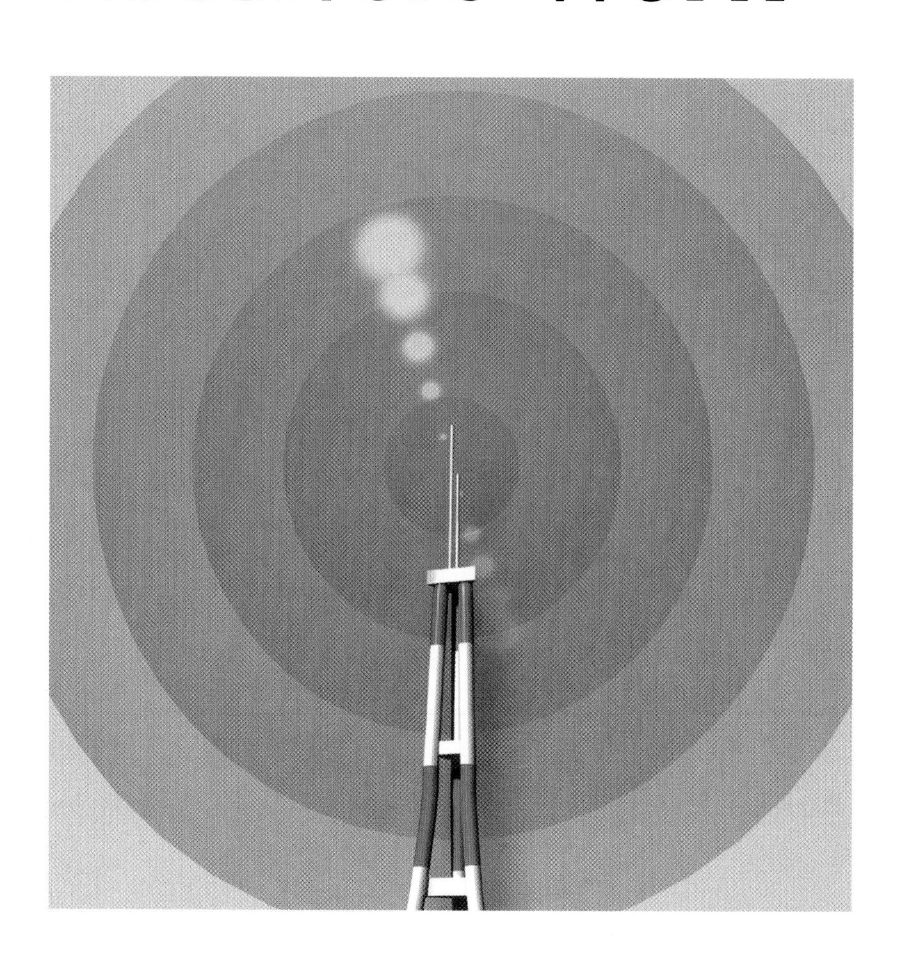

NO matter what kind of information is being sent and received, and no matter on what frequency it travels, the same basic hardware is required for all types of wireless technology.

This hardware must do several basic things. It must take information, such as music, that is originally an electrical signal, put that signal onto a carrier electrical signal, and then convert the electrical signal into an RF signal. Then, it needs to transmit that signal. On the receiving end, it must receive the signal, convert the RF wave into an electrical signal, separate the information from the carrier wave, and then interpret the resulting electrical signal in some way, such as by sending it into a headphone or speakers.

Obviously, in the real world things are more complicated than this, but these are the basic steps.

These steps are handled by three primary pieces of hardware: a transmitter, an antenna, and a receiver. The transmitter does the work of taking the information and piggybacking it onto another signal. The antenna does the job of converting that electrical signal to RF waves, which then propagate through the air. An antenna receives the waves, and then the receiver turns the waves into electrical energy, separates the information, and sends it to devices such as speakers or headphones.

In some instances, transmitters and receivers are found on the same device—such as a cell phone, which needs to both send information and receive it. In other instances, transmitters and receivers are separate—for example, in the instance of TV transmitters and TV sets. When a device does both sending and receiving, it's called a *transceiver*, and typically has one antenna that both sends and receives RF waves.

In some devices, such as a cell phone, transmitters and receivers are designed to send and receive only at a single frequency. In essence, the capability to send and receive at that frequency is programmed into the hardware of the transmitters and receivers. In other devices, such as radios and televisions, the receiver has been designed to accept and interpret signals at a range of frequencies. These kinds of devices are a bit more complicated because they require the device to be capable of tuning into different frequencies. So, when you turn a radio dial, for example, you're telling the receiver to listen for waves only at a specific frequency and to ignore the rest.

How Antennas Work

Sending and receiving Antennas are used to both send and receive RF signals. When an antenna is used to send signals, it converts electrical current containing the signal into RF waves. The current is created by a transmitter and, as the current flows through the antenna and encounters resistance, it creates the RF waves, which radiate outward.

Receiving When an antenna is used for receiving signals, it works in the opposite way as one used for sending. It receives RF waves and converts them into an electrical current containing the signal. Because the signal can be weak, some antennas contain a preamplifier that strengthens the signal before sending it on to the receiver.

Transmitter

Receiver

Omnidirectional and directional There are two general types of antennas: omnidirectional antennas and directional antennas. Omnidirectional antennas send out signals in all directions, whereas directional antennas send it in a specific direction. Directional antennas are used for many purposes; for example, when an antenna has a hill or mountain in back of it. A directional antenna will tend to send its signal farther, because it takes the energy that otherwise would be sent in all directions and concentrates it in one direction.

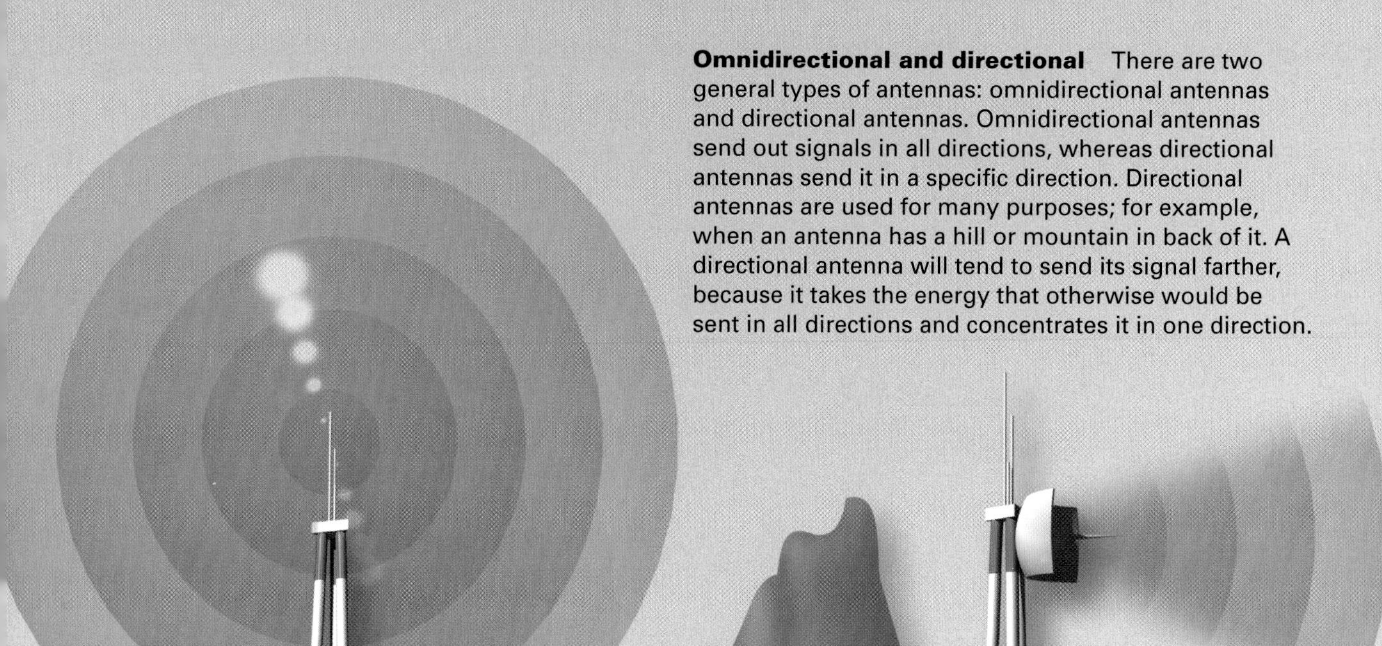

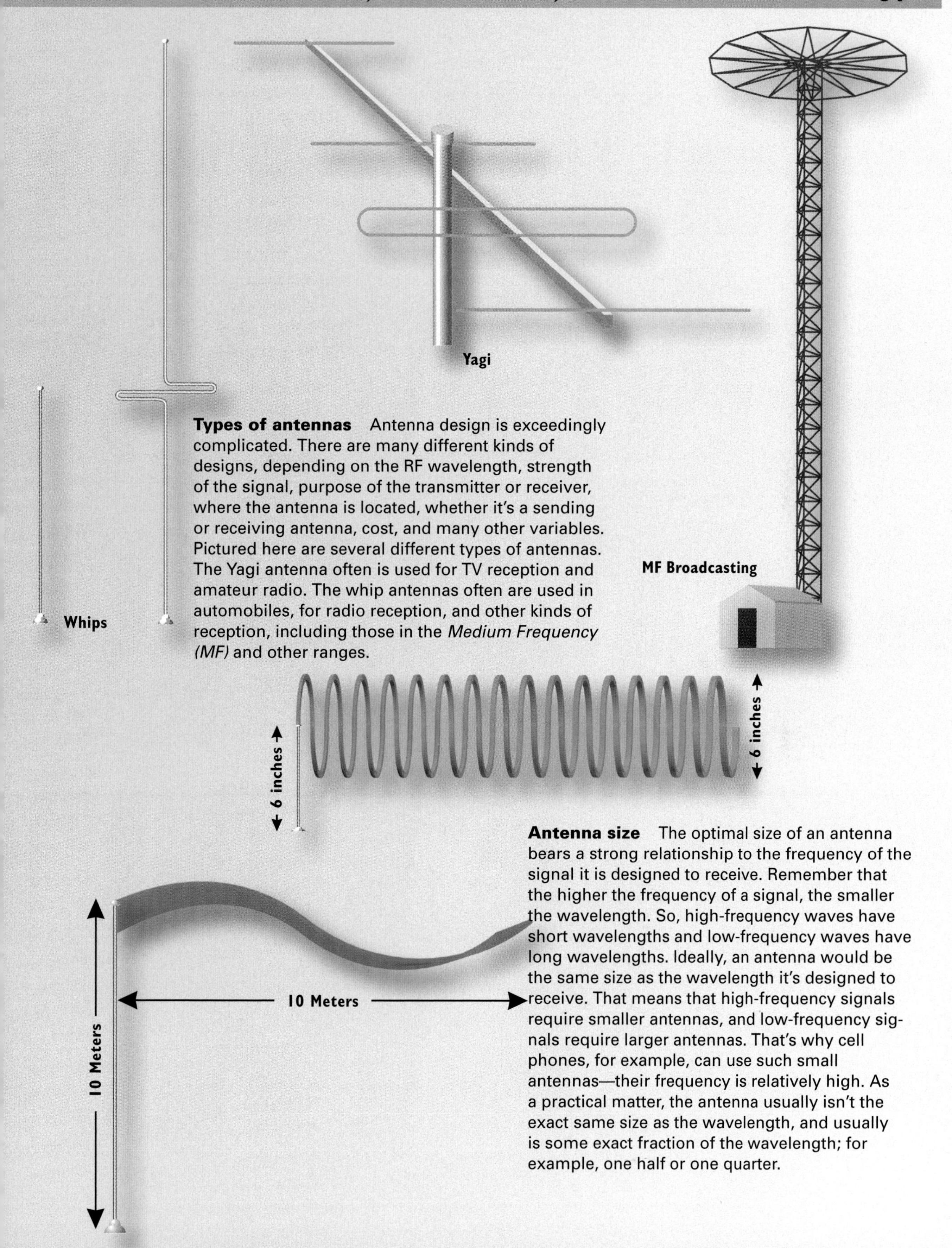

Types of antennas Antenna design is exceedingly complicated. There are many different kinds of designs, depending on the RF wavelength, strength of the signal, purpose of the transmitter or receiver, where the antenna is located, whether it's a sending or receiving antenna, cost, and many other variables. Pictured here are several different types of antennas. The Yagi antenna often is used for TV reception and amateur radio. The whip antennas often are used in automobiles, for radio reception, and other kinds of reception, including those in the *Medium Frequency (MF)* and other ranges.

Antenna size The optimal size of an antenna bears a strong relationship to the frequency of the signal it is designed to receive. Remember that the higher the frequency of a signal, the smaller the wavelength. So, high-frequency waves have short wavelengths and low-frequency waves have long wavelengths. Ideally, an antenna would be the same size as the wavelength it's designed to receive. That means that high-frequency signals require smaller antennas, and low-frequency signals require larger antennas. That's why cell phones, for example, can use such small antennas—their frequency is relatively high. As a practical matter, the antenna usually isn't the exact same size as the wavelength, and usually is some exact fraction of the wavelength; for example, one half or one quarter.

How Transmitters Work

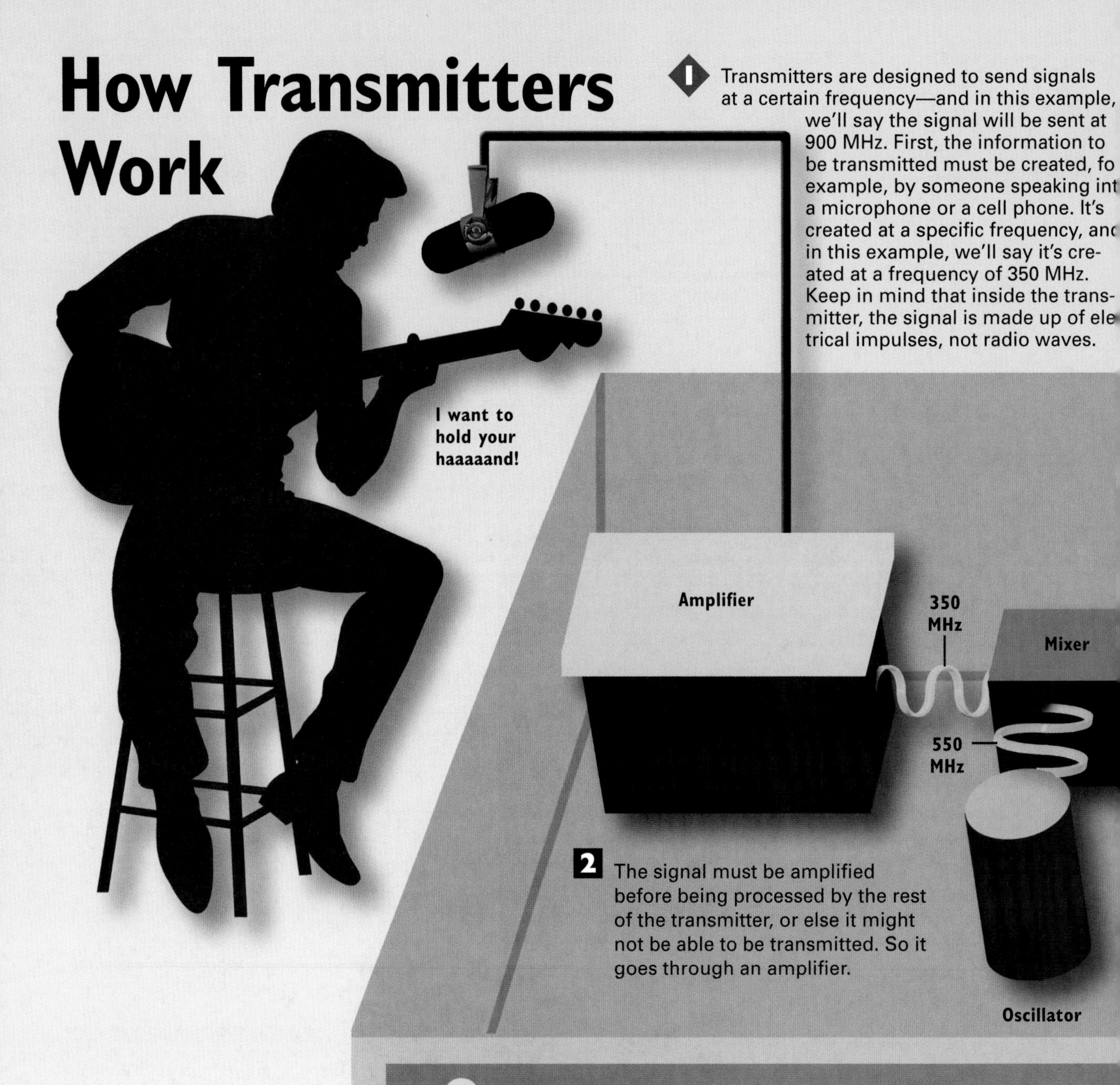

1 Transmitters are designed to send signals at a certain frequency—and in this example, we'll say the signal will be sent at 900 MHz. First, the information to be transmitted must be created, fo example, by someone speaking int a microphone or a cell phone. It's created at a specific frequency, anc in this example, we'll say it's created at a frequency of 350 MHz. Keep in mind that inside the transmitter, the signal is made up of ele trical impulses, not radio waves.

2 The signal must be amplified before being processed by the rest of the transmitter, or else it might not be able to be transmitted. So it goes through an amplifier.

3 The signal—in our example, the human voice—must be put onto a carrier wave to be transmitted. In our example, the frequency of the resulting carrier wave plus signal needs to be 900 MHz. So, a carrier wave needs to be created. The oscillator creates a carrier wave—it's designed to create as perfect a wave as possible. In our example, the frequency of the signal wave is 350 MHz, and the frequency of the wave being sent by the transmitter needs to be 900 MHz, so the oscillator needs to create a perfect wave at 550 MHz.

4 Both the 550 MHz wave from the oscillator and the 350 MHz wave from the amplifier are sent to a mixer, which takes the two waves and combines them. Waves at two frequencies come out of the mixer—one that is the sum of the frequencies, and another that is the difference between the two frequencies. In our example, that means that waves come out at both 900 MHz (550 MHz plus 350 MHz) and 200 MHz (550 MHz minus 350 MHz).

7 The signal now is ready to be transmitted. To be transmitted, its electrical signal must be converted into an RF wave. It travels to the antenna, which converts the electrical signal into an RF wave and sends it.

5 Before the signal can be transmitted, it must be cleaned of any waves at unwanted frequencies. Our transmitter is designed to transmit at 900 MHz, so there needs to be some way to get rid of the 200 MHz wave coming out of the mixer. The signals are sent through a filter, which gets rid of the unwanted signal; in this case, the 200 MHz one. There are four general types of filters. *Low-pass* filters allow any frequency below a certain frequency to pass through and eliminate the others. *High-pass* filters allow any frequency above a certain frequency to pass through and eliminate the others. *Bandpass* filters allow any frequency between two frequencies to pass through and reject all others. And *band reject* filters allow any frequencies except those found between two frequencies to pass through and reject all others.

Antenna

900 MHz

Filter

High-Power Amplifier

200 MHz

900 MHz

6 You now have a clear signal to transmit, but the signal is too weak at this point to travel far. So, the signal goes into another amplifier, a much more powerful one than the original one. Amplifiers in transmitters are called *high-power amplifiers (HPA)* because they're designed to boost the signal strength as much as possible. The amount of boost required varies according to the device and how far the signal needs to travel. A cell phone's base station, for example, has an amplifier 50 times as strong as that of a cell phone.

Wireless Tidbit

One of the reasons why filters are needed in transmitters is a legal one. The *Federal Communications Commission (FCC)*, which regulates the U.S. airwaves, requires by law that when a company is allowed to transmit at a given frequency, it can't transmit at any other frequency. That's because if it transmitted at other frequencies, it could interfere with other signals.

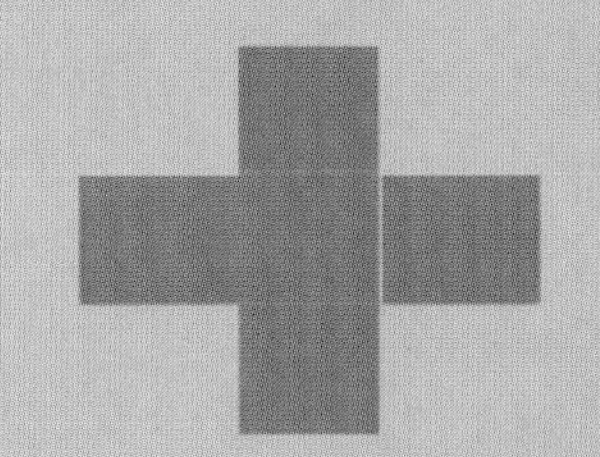

Transmitter

How Receivers Work

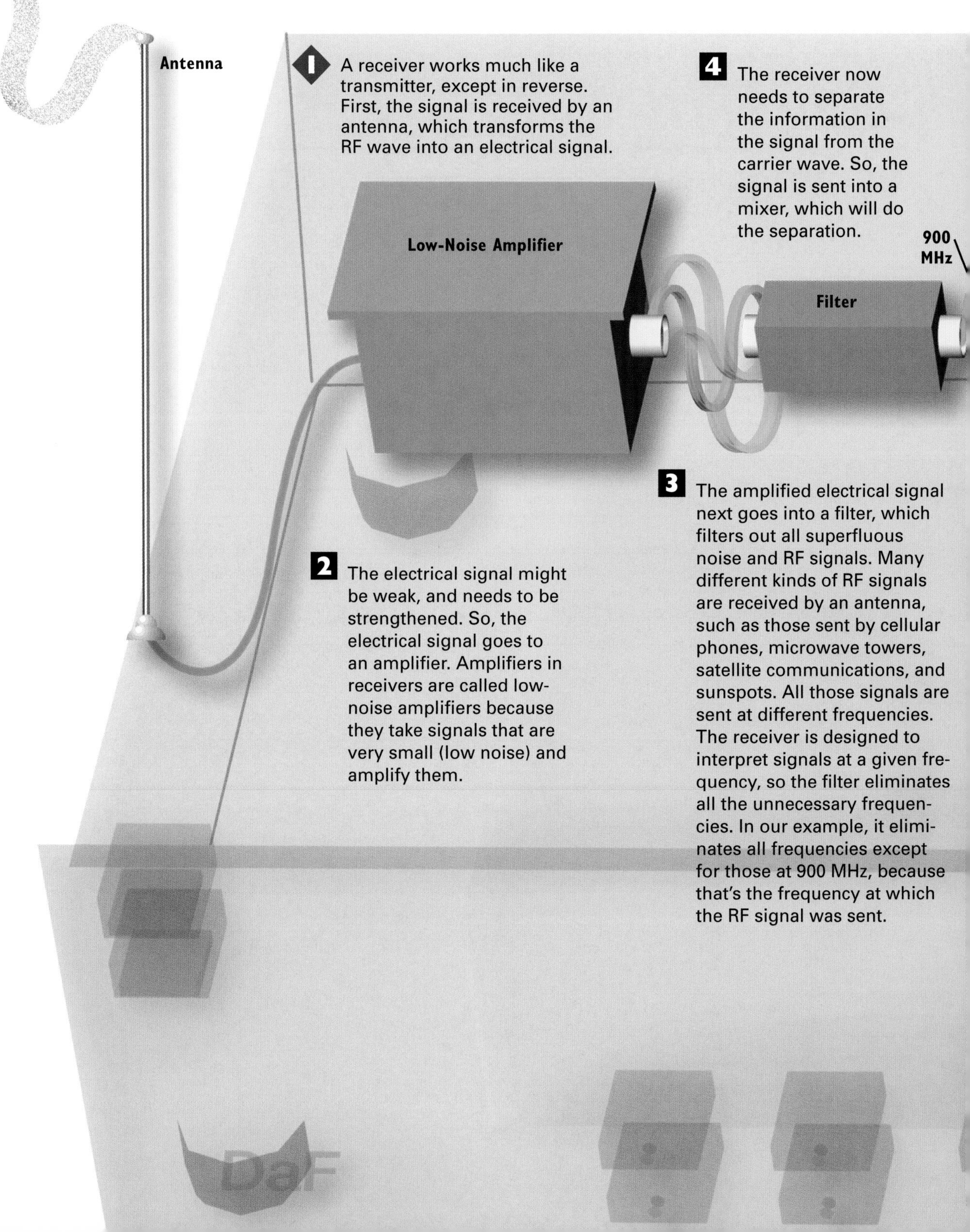

1 A receiver works much like a transmitter, except in reverse. First, the signal is received by an antenna, which transforms the RF wave into an electrical signal.

2 The electrical signal might be weak, and needs to be strengthened. So, the electrical signal goes to an amplifier. Amplifiers in receivers are called low-noise amplifiers because they take signals that are very small (low noise) and amplify them.

3 The amplified electrical signal next goes into a filter, which filters out all superfluous noise and RF signals. Many different kinds of RF signals are received by an antenna, such as those sent by cellular phones, microwave towers, satellite communications, and sunspots. All those signals are sent at different frequencies. The receiver is designed to interpret signals at a given frequency, so the filter eliminates all the unnecessary frequencies. In our example, it eliminates all frequencies except for those at 900 MHz, because that's the frequency at which the RF signal was sent.

4 The receiver now needs to separate the information in the signal from the carrier wave. So, the signal is sent into a mixer, which will do the separation.

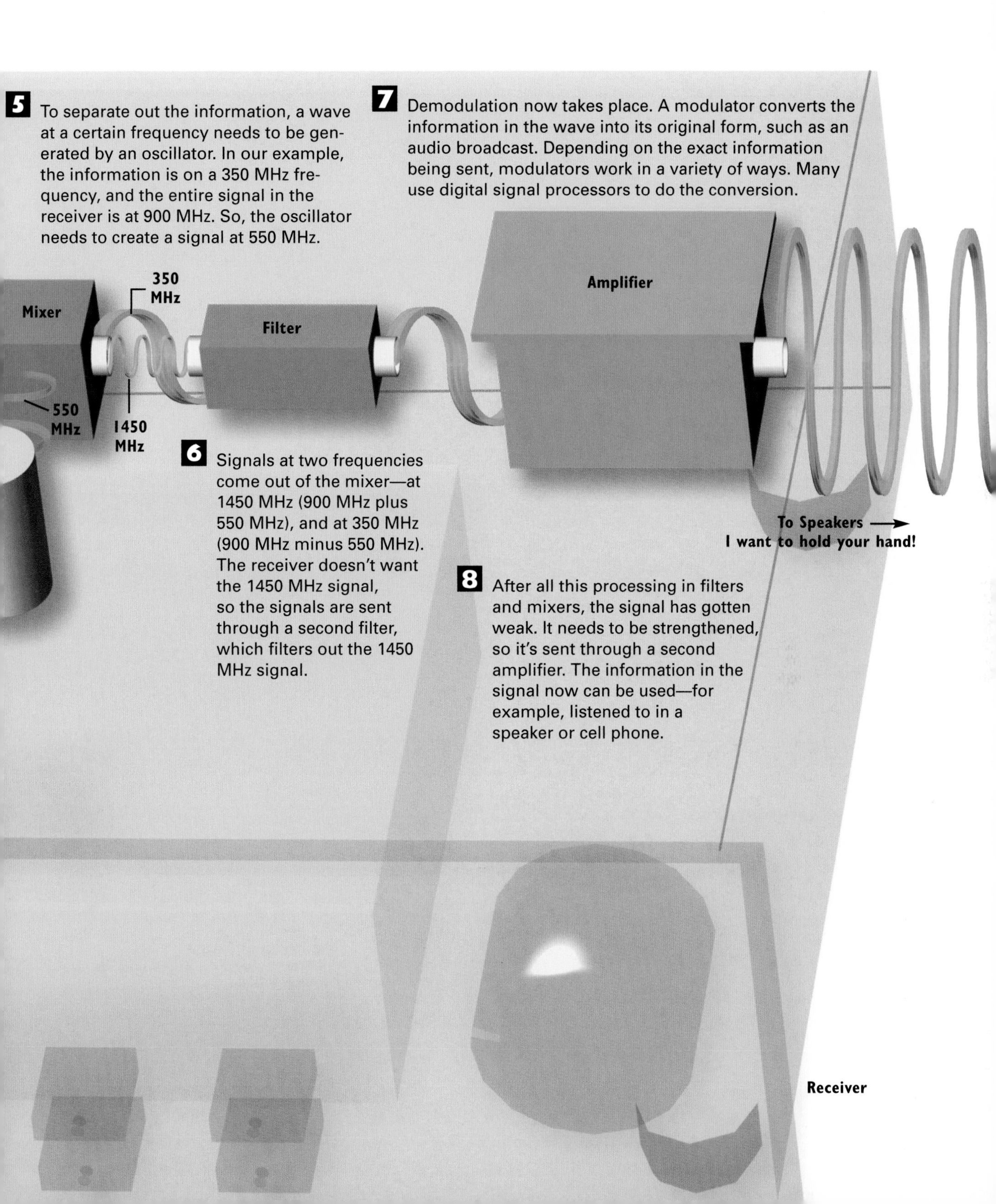
5 To separate out the information, a wave at a certain frequency needs to be generated by an oscillator. In our example, the information is on a 350 MHz frequency, and the entire signal in the receiver is at 900 MHz. So, the oscillator needs to create a signal at 550 MHz.
6 Signals at two frequencies come out of the mixer—at 1450 MHz (900 MHz plus 550 MHz), and at 350 MHz (900 MHz minus 550 MHz). The receiver doesn't want the 1450 MHz signal, so the signals are sent through a second filter, which filters out the 1450 MHz signal.
7 Demodulation now takes place. A modulator converts the information in the wave into its original form, such as an audio broadcast. Depending on the exact information being sent, modulators work in a variety of ways. Many use digital signal processors to do the conversion.
8 After all this processing in filters and mixers, the signal has gotten weak. It needs to be strengthened, so it's sent through a second amplifier. The information in the signal now can be used—for example, listened to in a speaker or cell phone.
350 MHz
Mixer
Filter
Amplifier
550 MHz
1450 MHz
To Speakers
I want to hold your hand!
Receiver

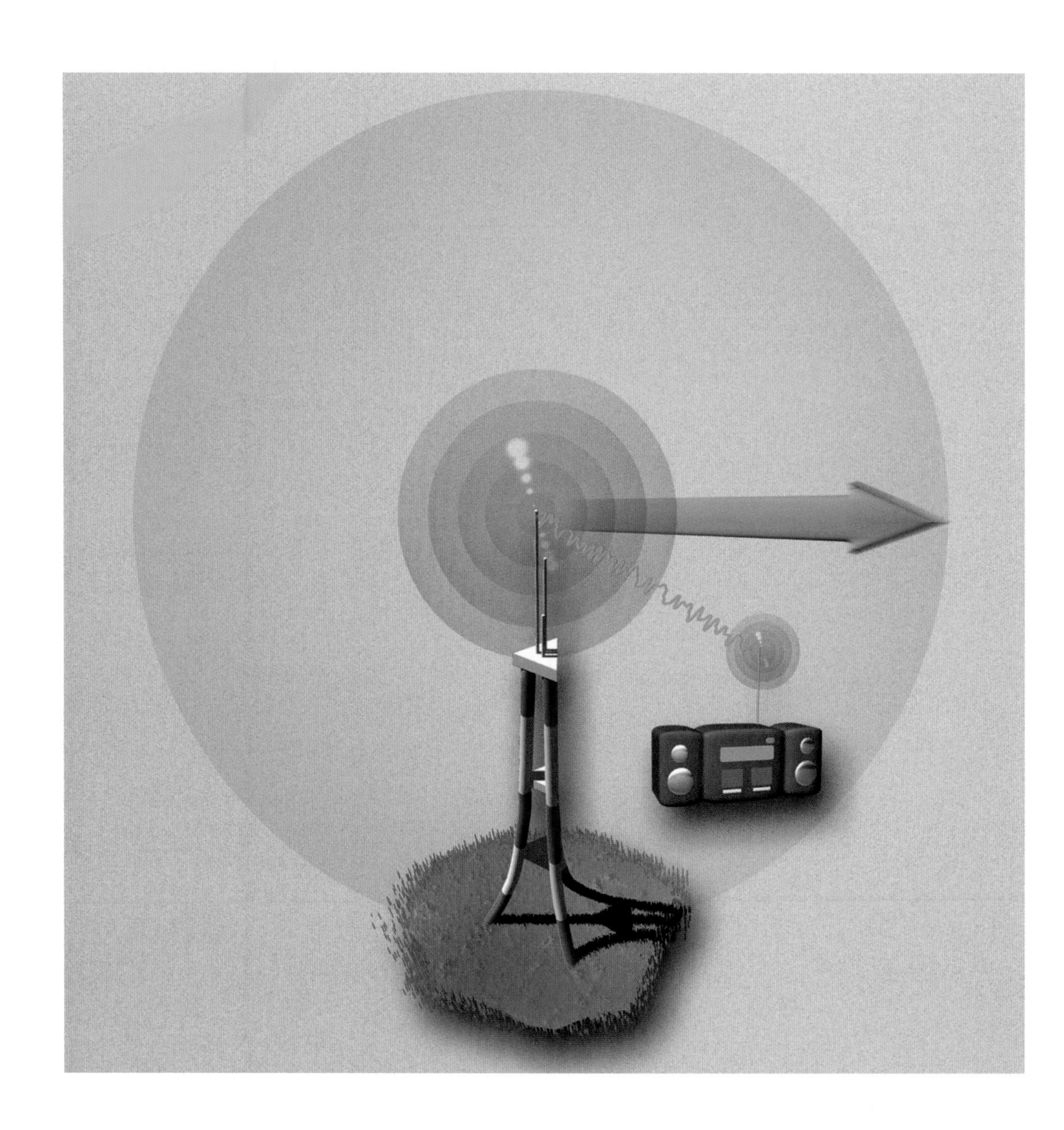

PART

How Radio and Television Work

WE like to think that the latest technologies are the most revolutionary the world has seen. The Internet has been hailed as the technology that finally will bring the world together in one vast, global village, and allows ways for people to communicate as never before—one on one across space, instantly. And it's been recognized as bringing instantaneous news to everywhere on the globe—free information available whenever people want, updated constantly throughout the day.

Cell phones and cellular technology are seen as similarly revolutionary. Wherever you are, you can talk with anyone else, untethered by wires.

The truth is, though, that the impact of the Internet and cellular technologies pales in comparison with technologies from the earliest days of the 20th century, and even from the latter days of the 19th century. Radio and television revolutionized the way that people work and think in ways far more basic and far-reaching than the Internet and cell phones have.

Guglielmo Marconi first sent information over radio waves in 1895; by 1905, the first long-distance wireless distress signal, an SOS, was sent. Radio soon boomed, and by the first decade of the 20th century, radio stations were everywhere.

Radio was the first mass communications medium, delivering instant news and entertainment long distance and creating the modern mass market. People felt connected with the world in ways they never had before. In fact, an argument can be made that radio, more than politics or any other cause, was a prime driving force behind forging a national unity in the United States.

Television accelerated what radio had begun. The first experimental TV broadcasts were begun in 1924, and it took nearly 30 years for television to become a mass phenomenon. But when it became one, it did so with a vengeance. Not only has it become the primary way that people get their news and entertainment; in many ways, it shapes the way that people look and think about the world. And it does so on a global scale, not just on a national one. It's what has helped turn the world into a global village.

In this part of the book, we'll look at the technologies that make radio and television work.

Chapter 7, "How Radio Broadcasting Works," covers the technologies that make radio broadcasts work. We'll start by looking at the basics of radio broadcasting—how broadcasts are created, how the signals are piggybacked onto waves, amplified, and sent out over antennas, and then how radios tune in and decode the radio signals. It covers both AM and FM broadcasts.

The chapter also looks at several radio broadcast technologies that are less well-known, but will become increasingly popular. The first, low-power FM radio broadcasting (LPFM), allows nonprofit groups to broadcast in a small area, such as an individual neighborhood or town. These groups allow people to bypass the major corporations that, in essence, own the radio airwaves, and create community programming of their own—programming that is local, and that has points of view and voices that you otherwise might not hear on commercial radio. As you'll see in this chapter, there are ways to ensure that LPFM doesn't interfere with existing radio broadcasts.

The chapter also looks at subscription satellite radio. This service beams high-quality digital radio broadcasts down from satellites for a monthly subscription fee. Because satellites cover the entire United States, you'd be able to drive from one end of the country to the other without losing a radio signal.

Chapter 8, "How Walkie-Talkies and Family Radio Service (FRS) Work," explains the inner workings of the familiar walkie-talkie, the device that many of us know from childhood but that can be, in fact, far more powerful than the toys of our youth. Walkie-talkies allow direct communications between people, without having to use cell phone towers, pay monthly subscription fees, or pay for connect time. It's a generic term that describes a variety of personal two-way radios—radios that let people communicate directly with one another. Family Service Radio, which we'll look at in detail, is a new, powerful form of walkie-talkie that allows people and families to keep in touch when camping, bike riding, or just in the neighborhood.

Finally, Chapter 9, "How Television Broadcasting Works," looks at the technology that everyone claims to hate, but that they spend an inordinate amount of time watching. We'll see how TV cameras work, how TV signals are combined and processed, and how they're broadcast and received. We'll also look at how a television set is able to decode and display a television signal.

The chapter also looks at two newer technologies, satellite TV and digital TV. Satellite TV enables you to receive hundreds of TV stations beamed down to you by satellites circling the globe. And digital TV is the next generation of television—it uses an exceedingly high resolution and can even allow several television signals, as well as data, to ride along with television signals.

So, if you've ever wondered how the biggest mass communications media of our time work, this is the section of the book for you.

CHAPTER
7

How Radio Broadcasting Works

TODAY, we tend to think of the Internet as the greatest communications revolution the world has ever seen. But the truth is, it probably has not had as revolutionary impact on the world as did the radio.

Information was first piggybacked onto radio waves by Guglielmo Marconi in 1895, and the world hasn't been the same since. By 1905, the first wireless distress signal—an SOS—was sent using radio waves. Radio soon became a growth industry; so much so that by 1912 the U.S. Congress passed the first laws regulating public use of the airwaves. The radio was the first mass communications medium that drew the world closer together, that created a mass audience, that allowed news and information to travel great distances to huge audiences instantly.

Although today's electronic equipment is far more sophisticated than the earliest radios, and although the quality of the sound received is far, far superior, in the most basic ways, radio hasn't changed much since broadcasting first began. The same basic physics and technology still holds.

Like all other RF communications, radio broadcasts must be piggybacked onto carrier waves for them to be sent. They're piggybacked using *amplitude modulation (AM)* and *frequency modulation (FM)* techniques—hence the name AM and FM radio. AM signals are more prone to interference, so its quality is not as good. But AM signals traditionally have a longer wavelength than FM signals and travel a greater distance. Radio receivers include an antenna and a variety of electronics to tune into and demodulate signals, and then play them over speakers.

Although the basics of radio haven't changed that much over the decades, the past few years have seen some intriguing new developments. The first is *low-power FM (LPFM)* radio broadcasting. This allows non-profit institutions to run inexpensive radio stations that serve individual neighborhoods, cities, or regions. These radio stations operate at lower power—either 10 watts or 100 watts—and can broadcast their signal only for less than four miles. These stations require licenses from the Federal Communications Commission to operate, and the FCC makes sure they don't interfere with existing broadcasters.

The other new development is the advent of subscription satellite radio. It's a commercial service that uses satellites to deliver hundreds of high-quality, stereo radio broadcasts for a monthly fee. One of the big draws of the service is that when you drive, you'll never lose the signal. That's because when you leave the area that one satellite covers, you'll enter an area that another satellite covers. So, you could drive across the country and listen to the same radio station the entire time if you wanted.

How Radio Broadcasting Works

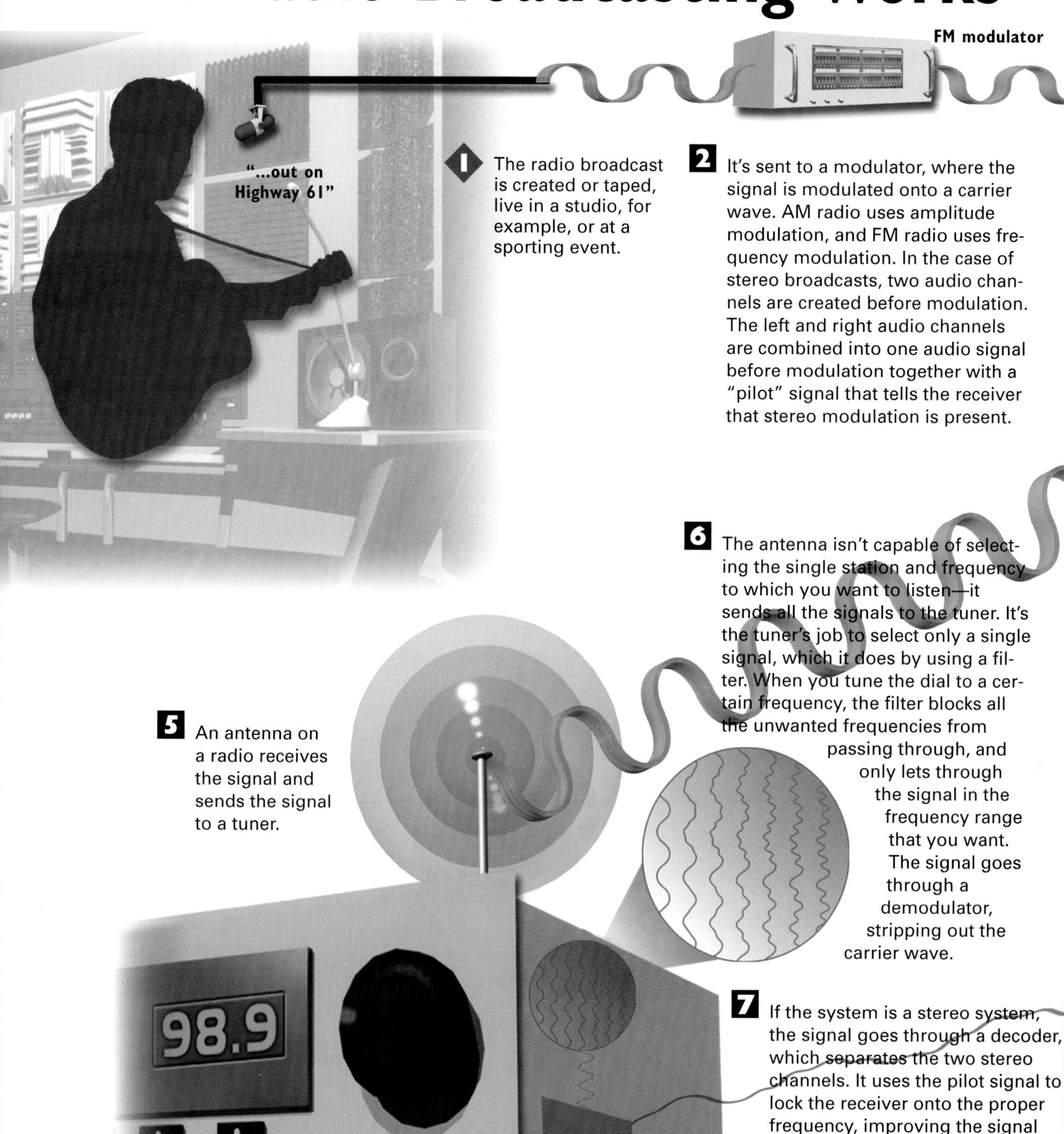

1 The radio broadcast is created or taped, live in a studio, for example, or at a sporting event.

2 It's sent to a modulator, where the signal is modulated onto a carrier wave. AM radio uses amplitude modulation, and FM radio uses frequency modulation. In the case of stereo broadcasts, two audio channels are created before modulation. The left and right audio channels are combined into one audio signal before modulation together with a "pilot" signal that tells the receiver that stereo modulation is present.

5 An antenna on a radio receives the signal and sends the signal to a tuner.

6 The antenna isn't capable of selecting the single station and frequency to which you want to listen—it sends all the signals to the tuner. It's the tuner's job to select only a single signal, which it does by using a filter. When you tune the dial to a certain frequency, the filter blocks all the unwanted frequencies from passing through, and only lets through the signal in the frequency range that you want. The signal goes through a demodulator, stripping out the carrier wave.

7 If the system is a stereo system, the signal goes through a decoder, which separates the two stereo channels. It uses the pilot signal to lock the receiver onto the proper frequency, improving the signal quality. The signal goes through an amplifier as well.

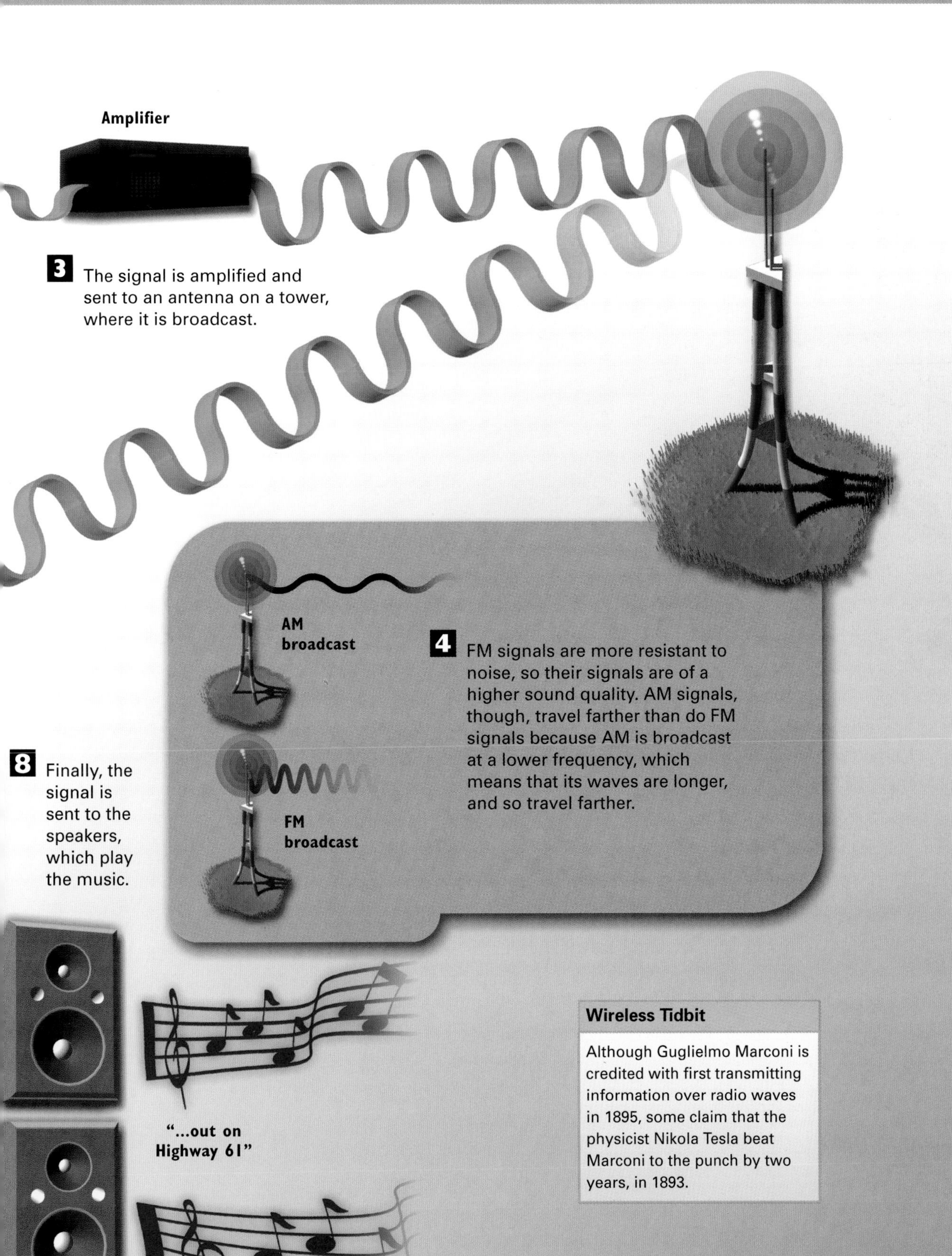

3 The signal is amplified and sent to an antenna on a tower, where it is broadcast.

4 FM signals are more resistant to noise, so their signals are of a higher sound quality. AM signals, though, travel farther than do FM signals because AM is broadcast at a lower frequency, which means that its waves are longer, and so travel farther.

8 Finally, the signal is sent to the speakers, which play the music.

Wireless Tidbit

Although Guglielmo Marconi is credited with first transmitting information over radio waves in 1895, some claim that the physicist Nikola Tesla beat Marconi to the punch by two years, in 1893.

How Low-Power FM Radio Broadcasting Works

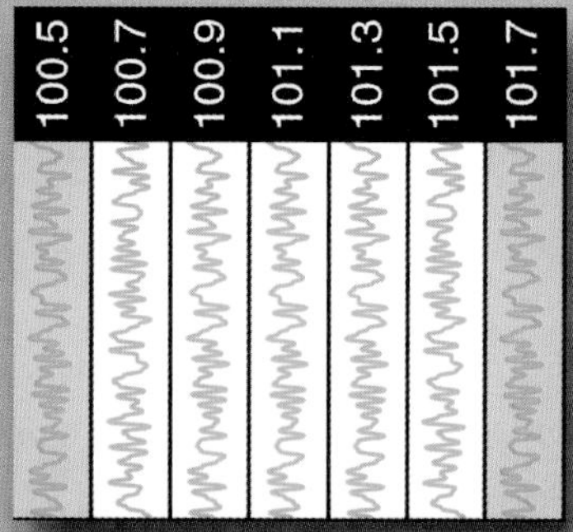

100.5 100.7 100.9 101.1 101.3 101.5 101.7

101.5

2 miles

1 *Low-power FM radio (LPFM)* allows nonprofit community groups, schools, churches, and similar organizations to run their own radio stations that broadcast within a small area the size of a neighborhood or small community. Groups that want to run a low-power radio station must apply to the *Federal Communications Commission (FCC)* for a license before they can set up a low-power radio station. The FCC requires that the station not interfere with existing FM radio stations. To ensure there is no interference, the frequency at which the station broadcasts must be separated from the frequencies of existing nearby stations by at least 400 MHz on each side. So, for example, if a low-power station wants to broadcast on 101.5 MHz, there must be no stations broadcasting at either 101.1 MHz or 101.9 MHz. Full-power radio stations must be separated by at least 600 MHz on each side.

4 People tune into low-power FM radio stations in precisely the same way they do other radio stations. There is no difference in the sound quality. The signal can just be heard only in a limited geographical area because it's weak.

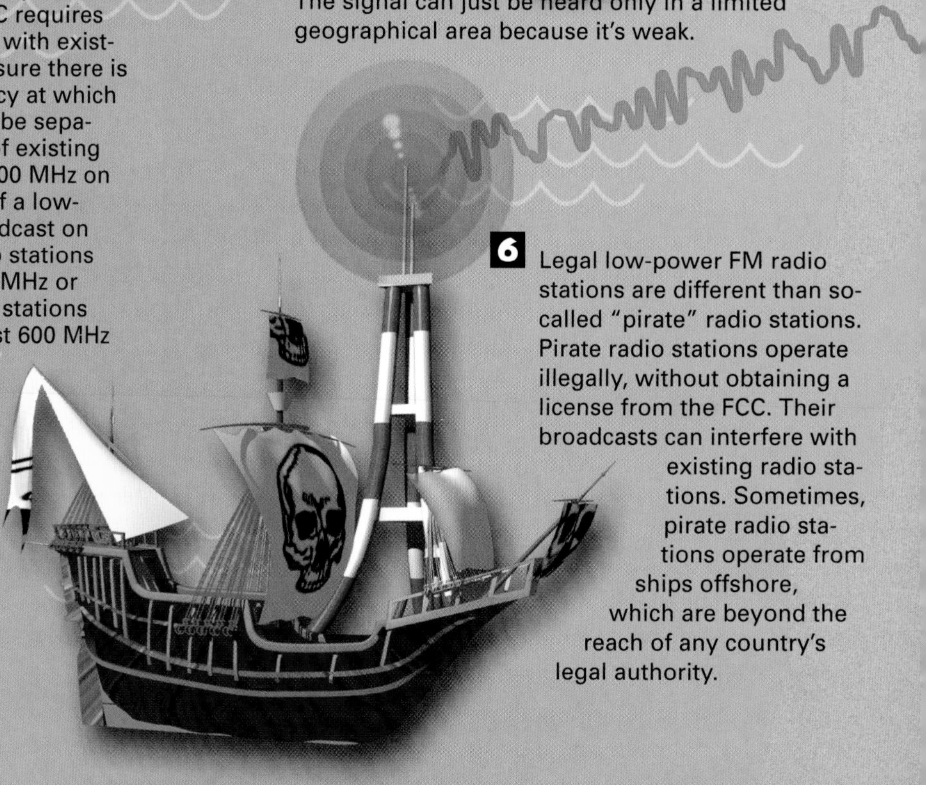

6 Legal low-power FM radio stations are different than so-called "pirate" radio stations. Pirate radio stations operate illegally, without obtaining a license from the FCC. Their broadcasts can interfere with existing radio stations. Sometimes, pirate radio stations operate from ships offshore, which are beyond the reach of any country's legal authority.

2 The radio station broadcasts like any other FM station—it sends signals from its studio to a radio tower and antenna, and the signal is broadcast from there. Because the radio station is broadcasting with low power, the equipment needed for it can be very inexpensive—as little as $3,000 to $5,000.

10 watts

2 miles

3 The radio station will be given a license to broadcast with a signal strength of either 10 or 100 watts. The signal from a 10-watt station can be heard up to approximately one to two miles from the antenna, whereas a 100-watt station can be heard up to approximately 3.5 miles from the antenna. By way of contrast, a 6,000-watt FM station can be heard approximately 18 miles from its antenna, and a 100,000-watt FM station can be heard approximately 60 miles from its antenna.

100 watts

3.5 miles

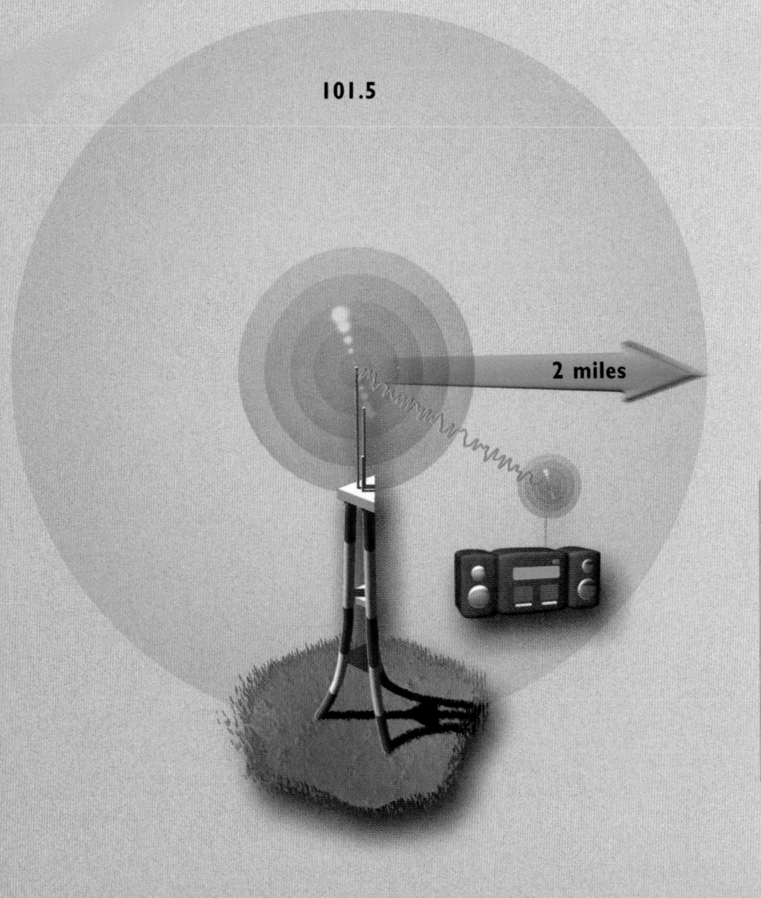

5 Because low-power FM radio stations broadcast only in a limited geographical area, many of them can be broadcasting in a given region, because they can use the same frequency without interfering with one another.

Wireless Tidbit

The Federal Communications Commission (FCC), the government agency that regulates the airwaves over which radio transmissions travel, including the radio and TV industries, was begun in 1927. It was first called the Federal Radio Commission, and was founded to sort out the chaos caused by the explosion of licensed and unlicensed radio stations in the U.S.

How Subscription Satellite Radio Works

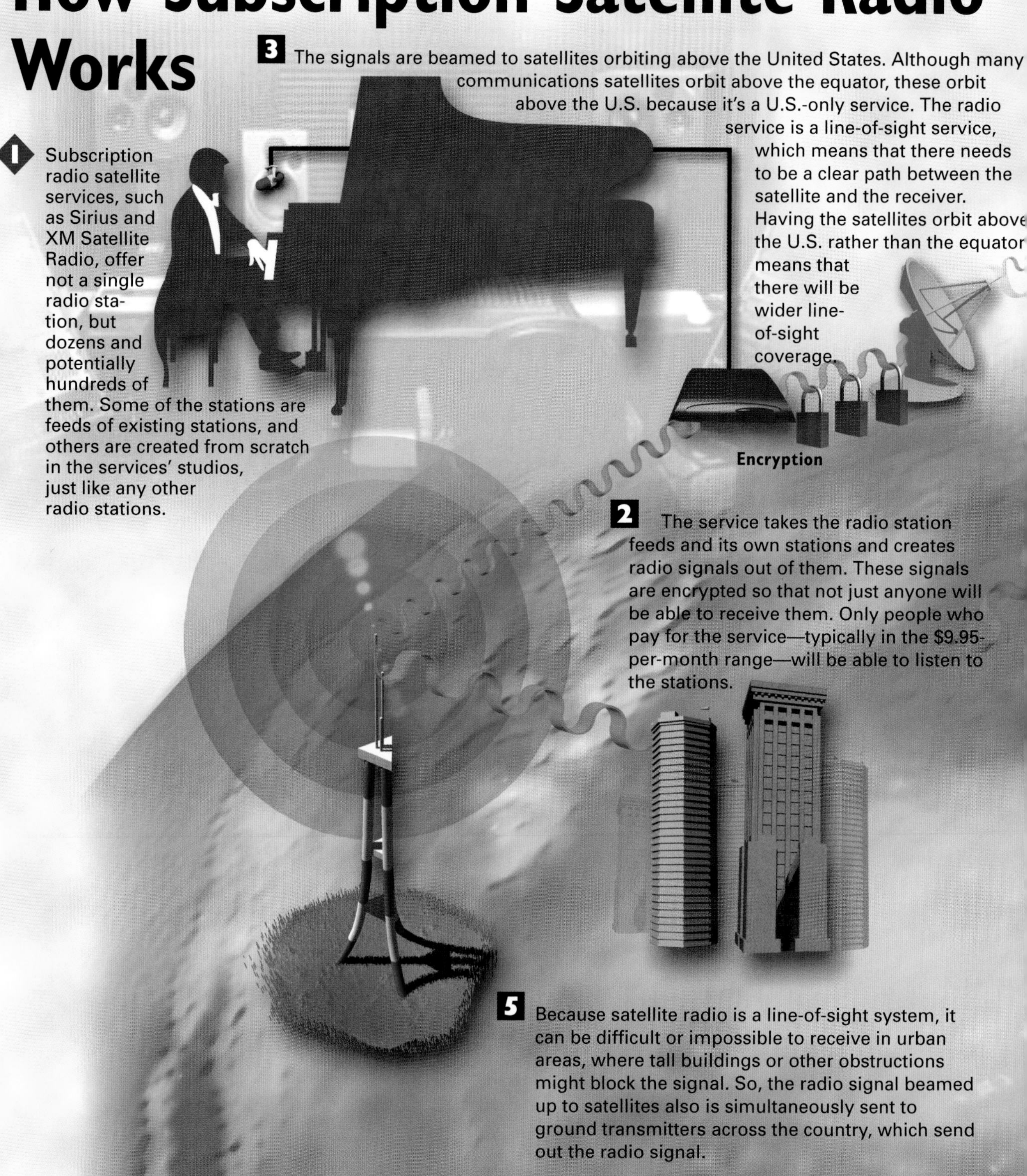

1 Subscription radio satellite services, such as Sirius and XM Satellite Radio, offer not a single radio station, but dozens and potentially hundreds of them. Some of the stations are feeds of existing stations, and others are created from scratch in the services' studios, just like any other radio stations.

2 The service takes the radio station feeds and its own stations and creates radio signals out of them. These signals are encrypted so that not just anyone will be able to receive them. Only people who pay for the service—typically in the $9.95-per-month range—will be able to listen to the stations.

3 The signals are beamed to satellites orbiting above the United States. Although many communications satellites orbit above the equator, these orbit above the U.S. because it's a U.S.-only service. The radio service is a line-of-sight service, which means that there needs to be a clear path between the satellite and the receiver. Having the satellites orbit above the U.S. rather than the equator means that there will be wider line-of-sight coverage.

5 Because satellite radio is a line-of-sight system, it can be difficult or impossible to receive in urban areas, where tall buildings or other obstructions might block the signal. So, the radio signal beamed up to satellites also is simultaneously sent to ground transmitters across the country, which send out the radio signal.

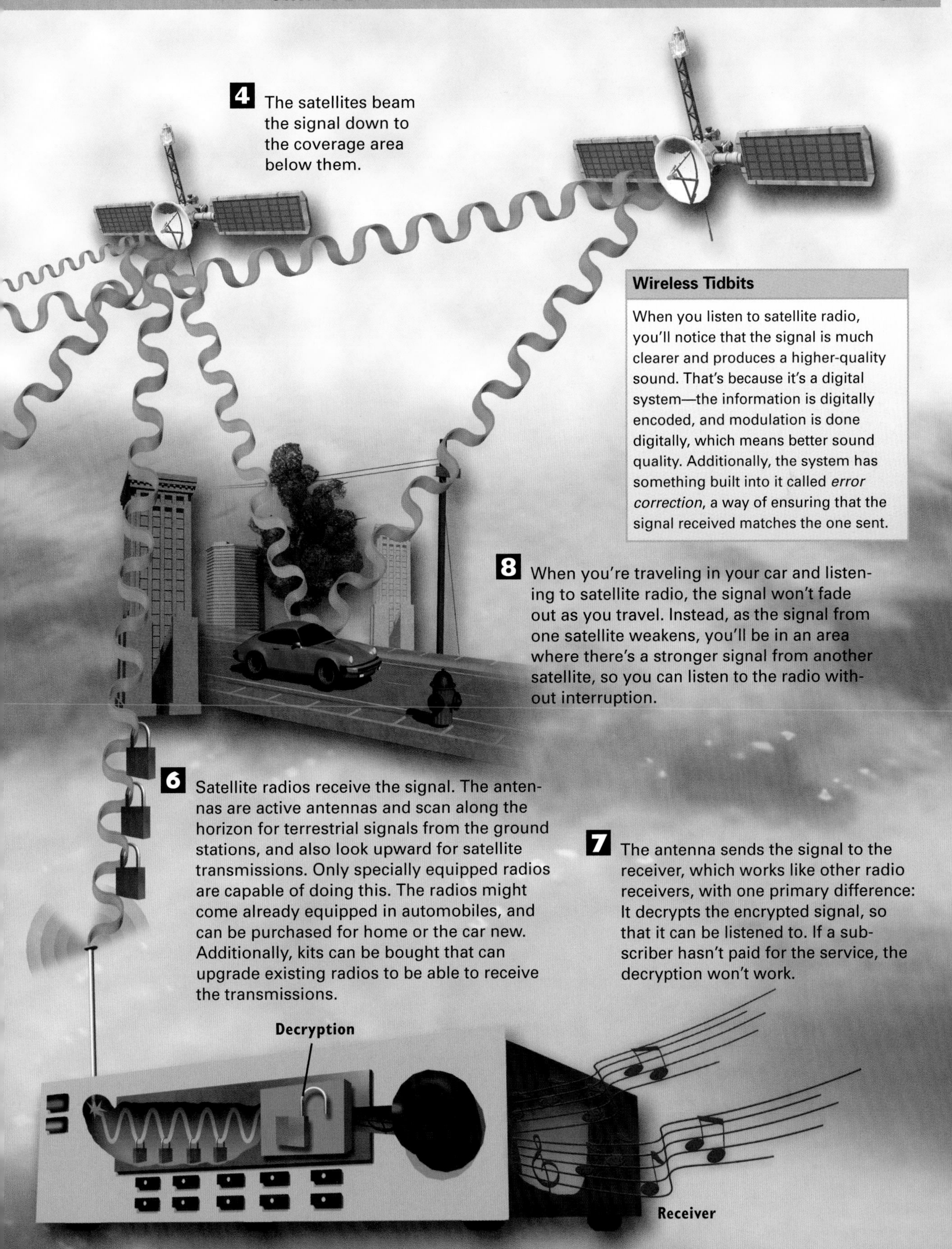

4 The satellites beam the signal down to the coverage area below them.

Wireless Tidbits

When you listen to satellite radio, you'll notice that the signal is much clearer and produces a higher-quality sound. That's because it's a digital system—the information is digitally encoded, and modulation is done digitally, which means better sound quality. Additionally, the system has something built into it called *error correction*, a way of ensuring that the signal received matches the one sent.

8 When you're traveling in your car and listening to satellite radio, the signal won't fade out as you travel. Instead, as the signal from one satellite weakens, you'll be in an area where there's a stronger signal from another satellite, so you can listen to the radio without interruption.

6 Satellite radios receive the signal. The antennas are active antennas and scan along the horizon for terrestrial signals from the ground stations, and also look upward for satellite transmissions. Only specially equipped radios are capable of doing this. The radios might come already equipped in automobiles, and can be purchased for home or the car new. Additionally, kits can be bought that can upgrade existing radios to be able to receive the transmissions.

7 The antenna sends the signal to the receiver, which works like other radio receivers, with one primary difference: It decrypts the encrypted signal, so that it can be listened to. If a subscriber hasn't paid for the service, the decryption won't work.

CHAPTER

8

How Walkie-Talkies and Family Radio Service (FRS) Work

WALKIE-TALKIES, also called handy talkies, have been with us for more than 60 years. Although the devices have been around since 1938, the way they work hasn't changed all that much over the years.

Walkie-talkies are transceivers—that is, they have hardware for both sending and receiving over RF. The same antenna both sends and receives. Walkie-talkies can't send and receive at the same time, however; Only one person can talk at a time when using them. Typically, when someone stops talking, you have to press a button to send a voice signal.

Walkie-talkie is, in fact, a generic term that describes a variety of personal *two-way radios*—radios that let people communicate directly with one another. Unlike cell phones and some other kinds of wireless communications technologies, walkie-talkies don't require separate transmitters or base stations. Instead, when you talk to someone, you transmit voice and they pick it up directly; and the same holds for when they transmit voice to you.

One of the most popular kinds of walkie-talkies became something of a cultural icon back in the 1970s—the CB (citizen's band) radio. CB radios communicate on shortwave frequencies at approximately 26 to 27 MHz. Their signals can travel long distances because they can "skip" through the atmosphere, but this skip also can cause problems—the radios have a problem communicating at more than five miles. But CB radios became a victim of their own success (and excess). They became so popular that foul language and general obnoxiousness began to take over the airwaves.

Other kinds of walkie-talkies are commercial walkie-talkies, used by companies in many different ways, such as for security personnel to communicate with one another, or at construction sites, or within a building.

There's one more type of walkie-talkie, what many people think of as the toys of their youth. These low-cost devices usually transmit only about 100 yards, and in fact, have no real use except for play.

Today, an increasingly popular kind of walkie-talkie is the *Family Radio Service (FRS)*. Think of FRS as the walkie-talkie of your youth on steroids, made more powerful by the use of FM as a means of modulation, and more power output. It can send and receive in a radius of up to two miles, and has a bevy of sophisticated features. You can use it to block out everyone except those in a select group. It will scan channels to find unused ones or to find people already talking. Some have scramblers to help keep your conversations private. The newest ones will send you weather alerts, and include Global Positioning Satellite (GPS) capabilities.

How Family Radio Service (FRS) Works

Channel 2 - 462.5875 MHz

460 MHz Band

1 *Family Radio Service (FRS)* is a two-way radio service that lets people talk to one another for no monthly fee, using a walkie-talkie–like FRS radio. No license is required to operate FRS. It operates in the 460 MHz band of the *Ultra High Frequency (UHF)* portion of the radio and transmits using FM. There are 14 channels on which it can communicate.

2 To communicate using FRS, you first must tune to one of the 14 channels on which it can communicate. The channels are 2.5 KHz apart—far enough apart so that there should be no interference between them.

CTSS Code 57

CTSS Code 57

5 Depending on where you are, all the channels can have many people on them. Using the *Continuous Tone Coded Squelch System (CTSS)*, you can block out the conversations of all the people on the channel except those you want to hear. Everyone in the group agrees to a one- or two-digit code and keys it into their radios.

4 Some FRS radios include scanners that can scan the channels and find which are occupied and which don't have people talking on them. That can help a group find a free channel on which they want to communicate.

Channel 2 - 462.5875 MHz

460 MHz Band

Find Channel

Wireless Tidbit

Walkie-talkies, CB radio, and pagers have more in common than you realize. They all were pioneered by the same remarkable inventor, Al Gross. Gross invented and patented the walkie-talkie in 1938, and then worked during World War II to develop a two-way ground-to-air communications system. In 1946, the FCC, following Gross's lead, allocated the first frequencies for private individuals to use personal radio, and dubbed them the Citizen Radio Service Frequency Band—the CB band. Gross formed a company whose equipment was the first to receive FCC approval for using the band, in 1948. The following year (1949), Gross patented the pager.

3 The person or group with which you're communicating must tune to the same channel. If they don't, you won't be able to talk to one another.

6 People in the group now will hear only other people in the group using the same code on the same channel. They won't hear any other conversations. However, the group's conversations are not private—everyone on the channel can hear what they say. CTSS only blocks the group from hearing what others say, but everyone else on the channel can hear what people in the group say.

CHAPTER

9

How Television Broadcasting Works

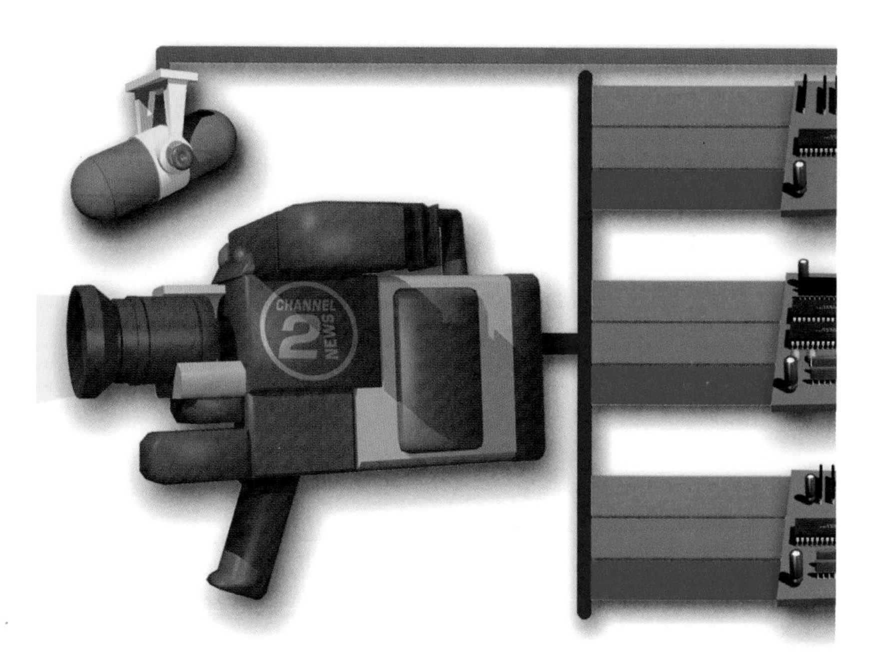

TELEVISION has been with us for nearly 80 years: The first experimental television broadcasts began in 1924. Fifteen years later, regularly scheduled broadcasts began, although in those early days very few people watched them. That shouldn't be a great surprise because the technology in the earliest days was very rudimentary—in TV's earliest incarnations, the picture was only one-inch square. TV pioneer Vladimir Kosma Zworykin, a Russian émigré to the United States, invented the cathode-ray tube, which enabled larger pictures and better transmissions. By the 1950s, the television set with its rabbit-ear antennas had become ubiquitous, and it took a central place in our culture—a place it has yet to relinquish.

If the world has truly become a global village, it has television to thank. News, complete with video, is beamed across the world even as it happens; many of the last several wars were in fact televised—and not just televised, but televised live. Increasingly, there is a global pop culture focused on MTV and Hollywood. More than any other wireless technology, it has conquered time and space.

Television broadcasts work like many other types of wireless communications, although with a variety of twists. First, the broadcast is created—and it might be live or on tape. The signal is processed, notably by separating it into red, green, and blue components, and then put onto a carrier RF signal and transmitted. Televisions pick up the signals with their antennas, decode and process the signal, and then display it by having rays of red, green, and blue strike the inside of the television set.

These days, of course, not all television is sent wirelessly; many people receive their TV signals through cable. But even for people who receive TV through cable, wireless transmission usually is involved. Cable systems often receive from satellites the signals they then send through the wires.

Although television is a nearly 80-year-old technology, there have been several advances in recent years, and some very big ones on the way. Many people prefer to receive their TV signals not through cable, but through satellite. To do that, people pay a subscription fee and get a small satellite dish that they must point toward a satellite. The satellite beams their signals down to them.

Satellite TV uses digital transmissions, which are of a higher quality than the analog signals now commonly used in normal broadcasts. But television is gradually going digital. Digital TV, and its offshoot High Definition TV (HDTV), have been slow coming along, but there's no doubt that they'll eventually arrive. HDTV allows for TV of an exceptionally high quality, including high-quality sound. And beyond that, because it's digital and uses computer formats for broadcasting, it will be easy for the signals to carry data as well, and will go a long way toward bringing interactivity to the TV set.

How Television Broadcasts Work

2 The three images go into a color mixer, which combines them to produce what's called a *luminance* signal. It gives the brightness of each part of the image.

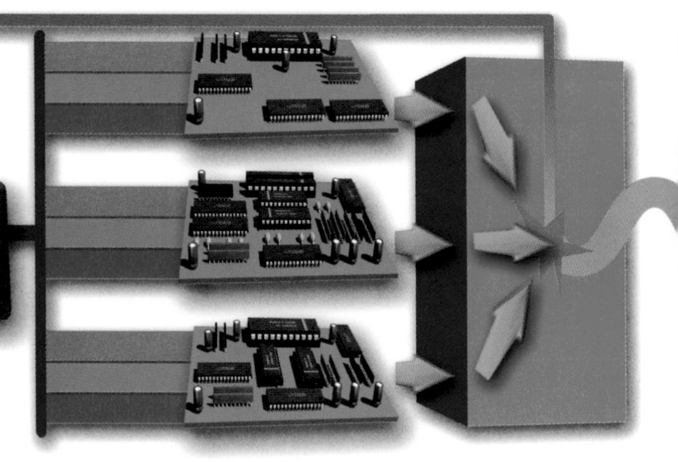

1 A TV camera separates the moving image into three images, one containing the red parts of the image, one containing the green parts of the image, and one containing the blue parts of the image. Together, these three colors are capable of reproducing any color in existence.

3 The images go into a color encoder, which creates what's called a *chrominance* signal. It details the amounts of each of the three different colors in each part of the image.

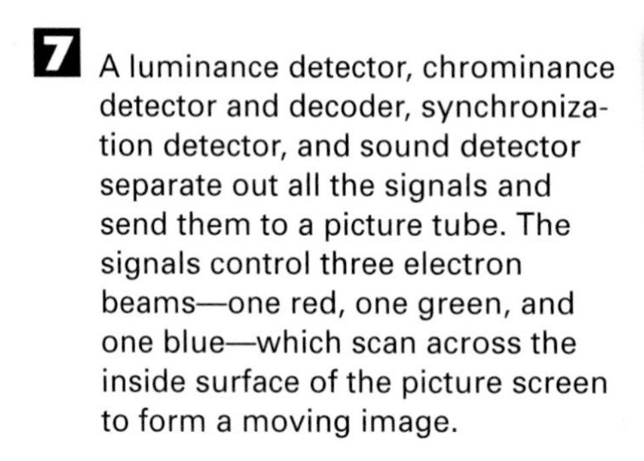

7 A luminance detector, chrominance detector and decoder, synchronization detector, and sound detector separate out all the signals and send them to a picture tube. The signals control three electron beams—one red, one green, and one blue—which scan across the inside surface of the picture screen to form a moving image.

4 A synchronization signal is added, which will ensure that when the moving image is displayed on a TV set, it will be properly displayed and synchronized. The chrominance, luminance, and synchronization signals are combined, and a sound signal is added.

Wireless Tidbit

The RF bands available for TV broadcasts only allow for the transmission of channels 2 through 83. Cable systems can deliver more than that number of channels because, in cable systems, the TV signal is sent to you over a cable rather than over the airwaves.

5 The signal is broadcast over RF waves, using a complex type of amplitude modulation called *Vestigial Sideband*. At the same time, the audio is sent through a separate FM signal transmitted alongside the video signal. A TV signal requires 6 MHz of bandwidth. TV broadcasts are sent in three bands of the RF spectrum: 54 to 88 MHz for channels 2 through 6; 174 MHz to 216 MHz for channels 7 through 13; and 470 to 890 MHz for channels 14 through 83. Each of those bands are cut into slices of 6 MHz of bandwidth for each channel.

6 The signal is received by an antenna, changed to an electrical signal, and sent through a tuner. When someone tunes to a particular channel, the tuner filters out all the unwanted frequencies and allows to pass only the frequency of the channel to which the person has tuned.

How Digital TV Works

1. *Digital TV (DTV)*, as its name implies, is an all-digital system that uses digital technology to send, receive, and play TV signals. There are several different standards for DTV, but the one getting the most attention is called *High-Definition TV (HDTV)*, which has the highest resolution of the DTV standards and also includes high-quality Dolby Digital surround sound. HDTV has a far higher resolution than existing analog TV—analog TV offers 535 lines of resolution, versus 720 or 1,080 lines of resolution, as you can get in HDTV.

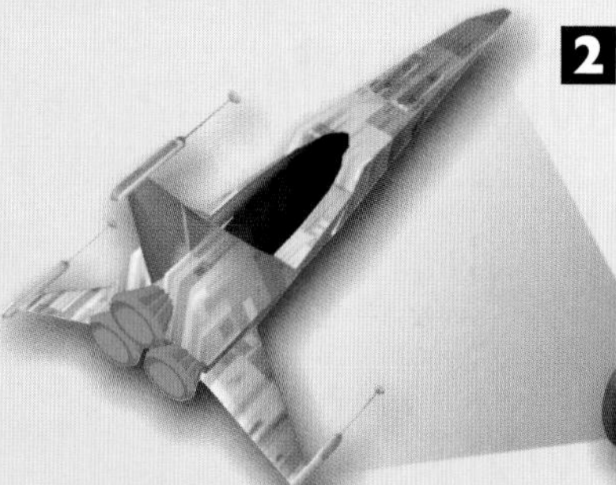

2. The higher resolution and digital sound means that a lot of information needs to be captured and broadcast. However, there isn't enough bandwidth devoted to each HDTV channel to broadcast all that information, so an HDTV signal first needs to be compressed before it's broadcast. The compression format is a computer compression format known as MPEG-2.

3. One of the ways MPEG-2 works is by recording only the changes to the image from a previous frame and recording the changes. For example, if a rocket is being launched and the background stays the same, it records only the motion of the rocket.

4 This and other techniques allow for a sizable amount of compression—it reduces the amount of data that needs to be transmitted by a ratio of 55 to 1, while still retaining an exceptionally high quality of signal, far better than in analog TV.

5 The compressed digital signal is broadcast. Broadcasters have been given channels with enough broadcast bandwidth to send the signal at 19.39 megabits of data per second. They can send a single program at the full 19.39 megabits per second, or can instead divide their channel into several subchannels and send several programs at not quite as high quality. Or, they can mix sending data information along with the video signal.

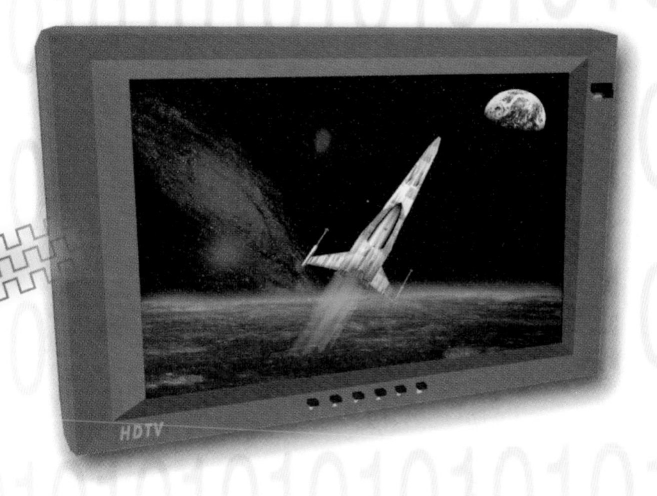

6 A special HDTV receiver and TV set are required to display HDTV signals. Not only is the signal of a higher resolution and the sounds of a better quality, but the image itself is much larger as well. Because HDTV televisions work by decoding MPEG-2 files, based on a computer format, CD-ROMs should be able to be played on them.

How Satellite TV Works

1 The most popular kind of systems for receiving TV through satellites use small antennas and are subscription services. In the systems, a television signal is created as it would be normally, as an analog signal. Satellite TV systems transmit signals digitally, so the signal goes through an analog-to-digital converter to convert the signal into a digital one. Note: If a signal is created as a digital signal, it does not need to go through this conversion.

2 To be sure that no one can steal the signal, it is scrambled so that only subscribers can view it.

Analog-to-Digital Converter

Scrambler

8 Inside the set-top box also is a descrambler that unscrambles the signal so that you can view it.

Wireless Tidbit

Small satellite dishes are popular today, but the first TV satellite dishes were exceedingly large and often were used in rural areas that didn't get many TV stations. To use these older, large dishes, you had to point the dish at a specific satellite.

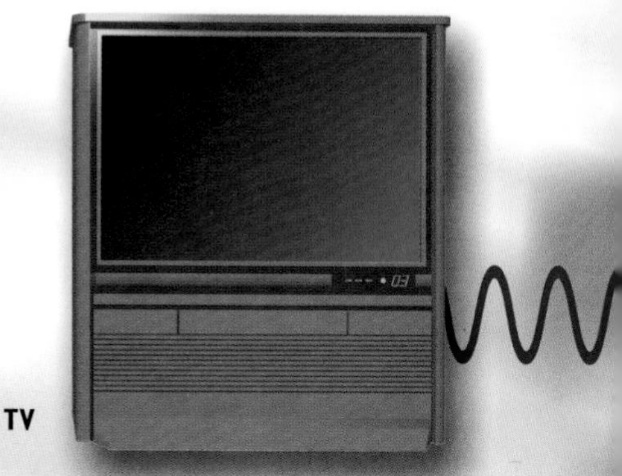

TV

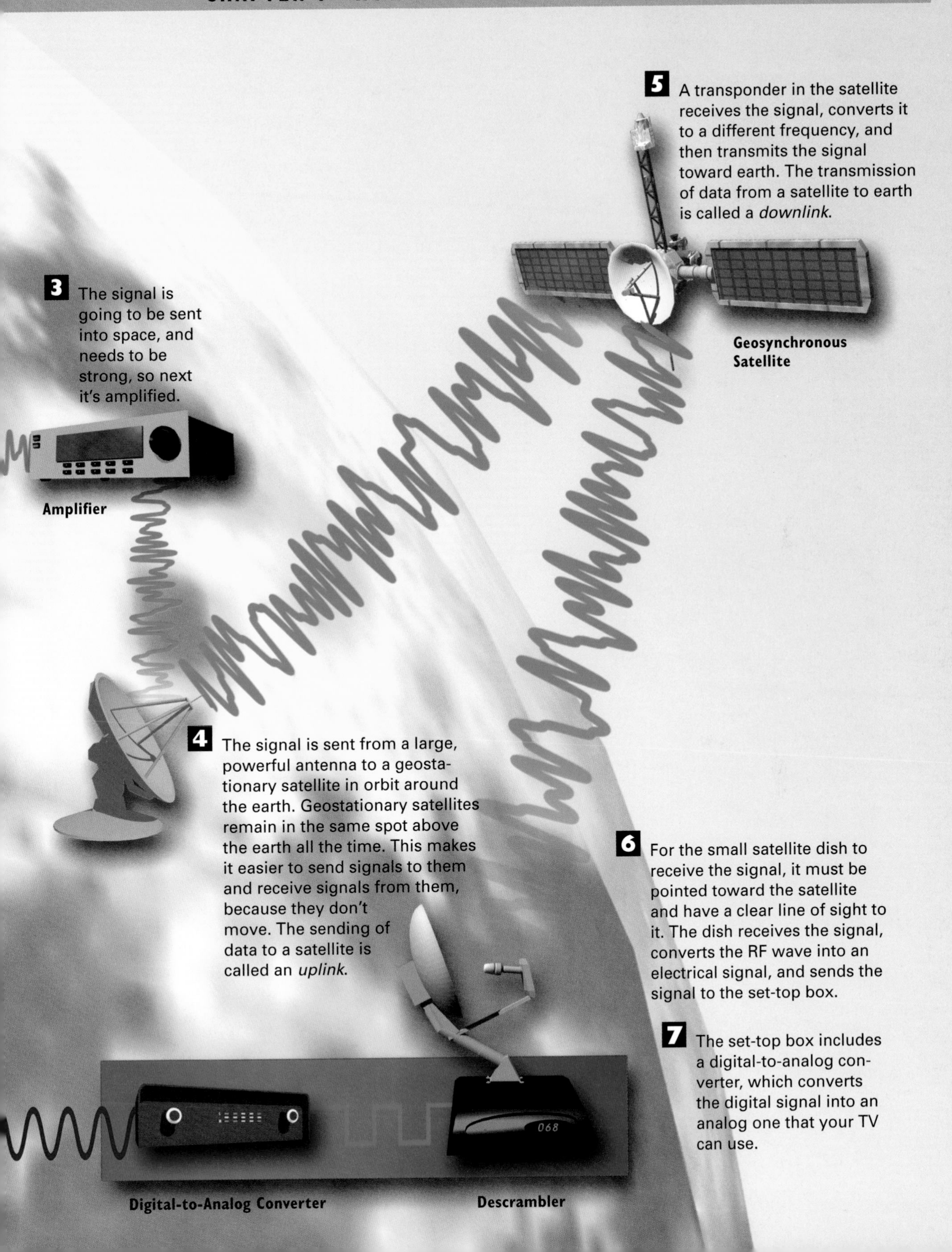

5 A transponder in the satellite receives the signal, converts it to a different frequency, and then transmits the signal toward earth. The transmission of data from a satellite to earth is called a *downlink*.

3 The signal is going to be sent into space, and needs to be strong, so next it's amplified.

4 The signal is sent from a large, powerful antenna to a geostationary satellite in orbit around the earth. Geostationary satellites remain in the same spot above the earth all the time. This makes it easier to send signals to them and receive signals from them, because they don't move. The sending of data to a satellite is called an *uplink*.

6 For the small satellite dish to receive the signal, it must be pointed toward the satellite and have a clear line of sight to it. The dish receives the signal, converts the RF wave into an electrical signal, and sends the signal to the set-top box.

7 The set-top box includes a digital-to-analog converter, which converts the digital signal into an analog one that your TV can use.

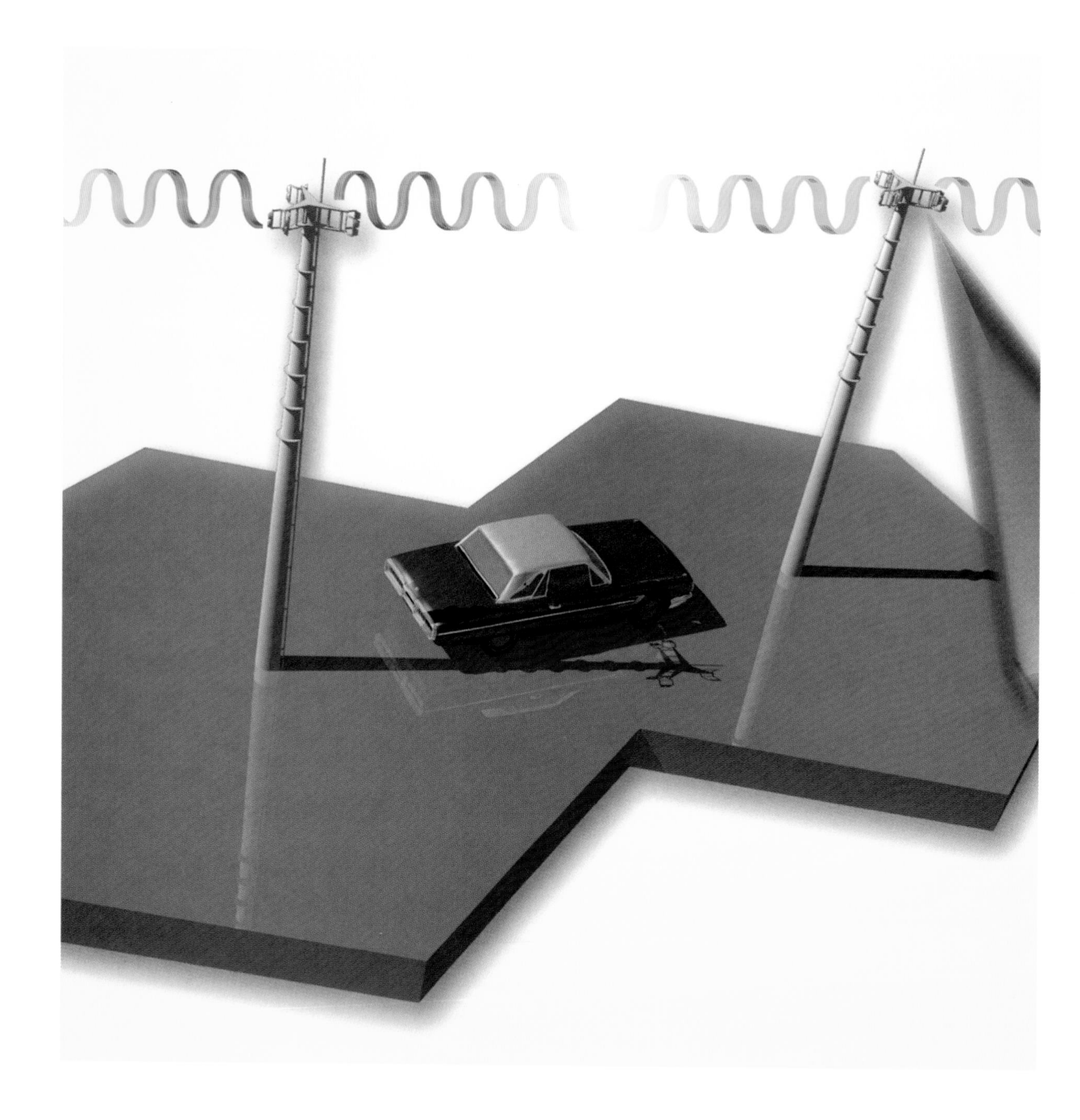

P A R T

How Cellular Telephones and Pagers Work

MENTION the word "wireless" to someone, and the first things they'll probably think of are cell phones or pagers. Forget radios, televisions, walkie-talkies, and the myriad other wireless devices out there. When it comes to wireless, people think cell phones.

There's good reason for that. They've become ubiquitous. You can't walk down a city street, drive in your car, or be anywhere else, for that matter, and not see someone with a cell phone held up against their ear. And it's not just human beings that have become so used to cell phones—some birds, notably starlings, have begun to copy the tones that cell phones make when they have a call.

As prevalent as cell phones are in the United States, they're that much more common in Asia and Europe. In fact, those continents are ahead of the United States not just in cell phone use, but also in the sophistication of their cell phone networks and features. Japan's i-mode, for example, offers the kind of interactive services that are still far away in the United States.

Pagers are common all over the world. They're not as noticeable because they're smaller. And increasingly, cell phones are replacing pagers. But pagers still are in widespread use, for everyone from plumbers to doctors.

In this part of the book, we'll take a look at how cell phones and pagers work.

Chapter 10, "How Cellular Networks, Cells, and Base Stations Work," explains all the basics of cell phones and their associated networks. You'll learn all about cellular networks—you'll see, for example, how a typical network routes a call from your phone, through base stations, switches, and a variety of other communications devices, and then sends those calls to the phone network or to other cellular networks. You'll see step-by-step the intricate choreography involved when you do a simple thing such as connect your phone to the network. Similarly, you'll see how your call is routed when you make a call, and how the network knows where you are and delivers a call to you when someone is trying to reach you.

The chapter also explains how individual cells work—the cells that give cell phones their name. Every cellular network is divided into many cells, each of which has its own transmitter and receiver called a base station. Cells cover a specific geographic area. You'll see how cells perform a "handoff" when you drive from one cell to another so that you can keep talking even though you're moving out of one cell and into another.

Chapter 11, "How Cellular Telephones Work," takes a closer look at the inner workings of the phone itself. In that chapter, you'll see a cutaway view of a cell phone, so that you can see all the different components and what each does to help make you send and receive calls.

You'll learn how cell phones use cellular *channels*—communication lines between the phone and the network that carry voice as well as commands that instruct the cell phone what to do to communicate with the network.

The chapter also explains the difference between digital and analog cell phones, and it spends a great deal of time making sense of the alphabet soup of cellular-related acronyms. You'll look inside a PCS system, you'll learn the difference between TDMA, CDMA, and GSM systems, and find out how they all work. And you'll learn how cell phones use the Short Message System (SMS) that allows people to send text messages to each other—much like a computer's instant messaging, except for cell phones.

Finally, this section of the book takes an inside look at pagers. Chapter 12, "How Pagers Work," shows you a cutaway view of a pager. You'll see its innards and how they all work together to deliver pages to you. You'll see how a paging network functions and what happens when someone calls in a page—you'll see it wend its way through the network, find out exactly where you are, and then page you.

The chapter also looks at an increasingly popular device—two-way pagers. These let you not only receive pages, but let you send messages as well. Although they have many uses, the most popular use is for getting always-on access to e-mail. With a two-way pager, you can always be alerted when you have e-mail, and can then read it and respond to it.

CHAPTER

10

How Cellular Networks, Cells, and Base Stations Work

CELL phone networks are everywhere, not just in the United States, but all over the world. In fact, cell phone use is more common and advanced outside the United States than inside it.

Analog networks, which were the first cellular networks, are referred to as *Advanced Mobile Phone Service*, or *AMPS*. They use frequency modulation to deliver signals. They were the first generation of cellular technology. But although AMPS is very popular, there are problems with analog networks like it. First, their capacity is limited, so they can't handle as many calls as more advanced cellular networks. And equally important is that they can't deliver the same kinds of advanced services, such as browsing the Web, paging, and text messaging, as digital networks.

Digital networks were developed to solve these problems. One of the most common one is called *Personal Communications Services (PCS)*. There are many other digital network schemes, though, as you'll learn in Chapter 11, "How Cellular Telephones Work," but no matter where cell phone networks are located, and whether they're digital or analog, they operate on similar principles. Those are the principles you'll learn about in the illustrations on the following pages.

Cell phone networks are made up of the phones themselves; of individual cells and their associated base stations, which communicate with the cell phones; and of a variety of networking hardware and software that handles internal communications for transferring calls and data through the network; and external communications for transferring calls and data from the network to other networks and to the normal telephone system. (The "normal" telephone system, by the way, is commonly known as the public switched telephone network, or *PSTN*.)

Cell phone networks are not monopolies, so more than one network can operate in a given geographical area. Depending on the type of network, they operate on different frequencies, so there isn't any interference between the networks.

As cell phone networks become ubiquitous, they raise certain societal issues. Whether drivers should be allowed to talk on their cell phones while driving has become a hot issue, and in fact, some municipalities have banned or curtailed the practice, and others are examining the issue. Some restaurants are banning cell phone use. And everyone has had the unpleasant experience of being in a movie theater, play, or opera and listening to someone's cell phone beep during important parts of the performance.

Of special concern to network operators, though, are that the cell phone towers which house cell base stations have begun to draw fire in certain communities. In particular, they've been attacked as being unsightly. Because of this, some cellular network operators have taken to disguising their towers as trees—designing them and camouflaging them to fit more closely into the natural environment so that they're not noticed by passing drivers or hikers. And some companies disguise their towers to look like flagpoles on the tops of buildings.

How Cellular Networks Work

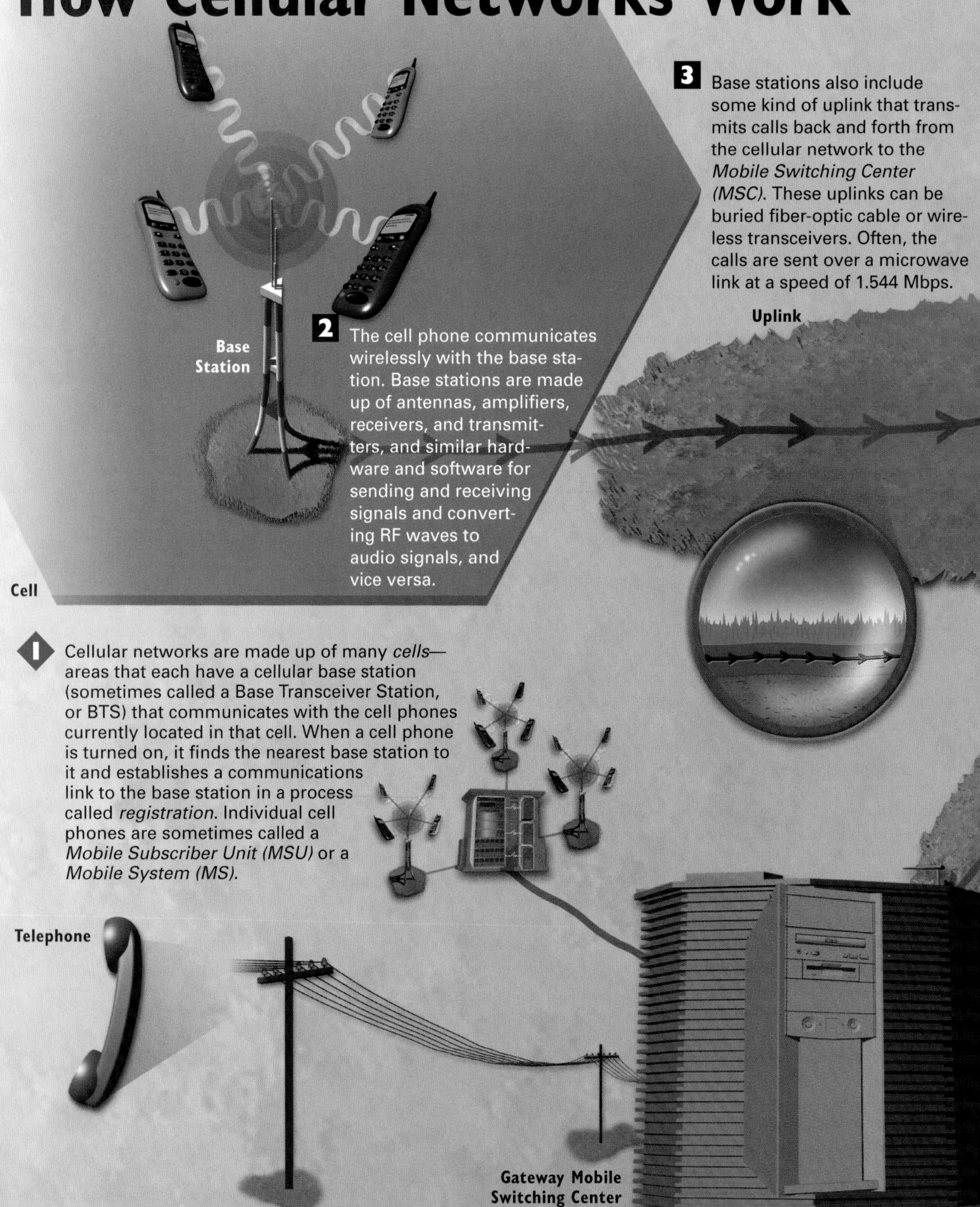

1 Cellular networks are made up of many *cells*—areas that each have a cellular base station (sometimes called a Base Transceiver Station, or BTS) that communicates with the cell phones currently located in that cell. When a cell phone is turned on, it finds the nearest base station to it and establishes a communications link to the base station in a process called *registration*. Individual cell phones are sometimes called a *Mobile Subscriber Unit (MSU)* or a *Mobile System (MS)*.

2 The cell phone communicates wirelessly with the base station. Base stations are made up of antennas, amplifiers, receivers, and transmitters, and similar hardware and software for sending and receiving signals and converting RF waves to audio signals, and vice versa.

3 Base stations also include some kind of uplink that transmits calls back and forth from the cellular network to the *Mobile Switching Center (MSC)*. These uplinks can be buried fiber-optic cable or wireless transceivers. Often, the calls are sent over a microwave link at a speed of 1.544 Mbps.

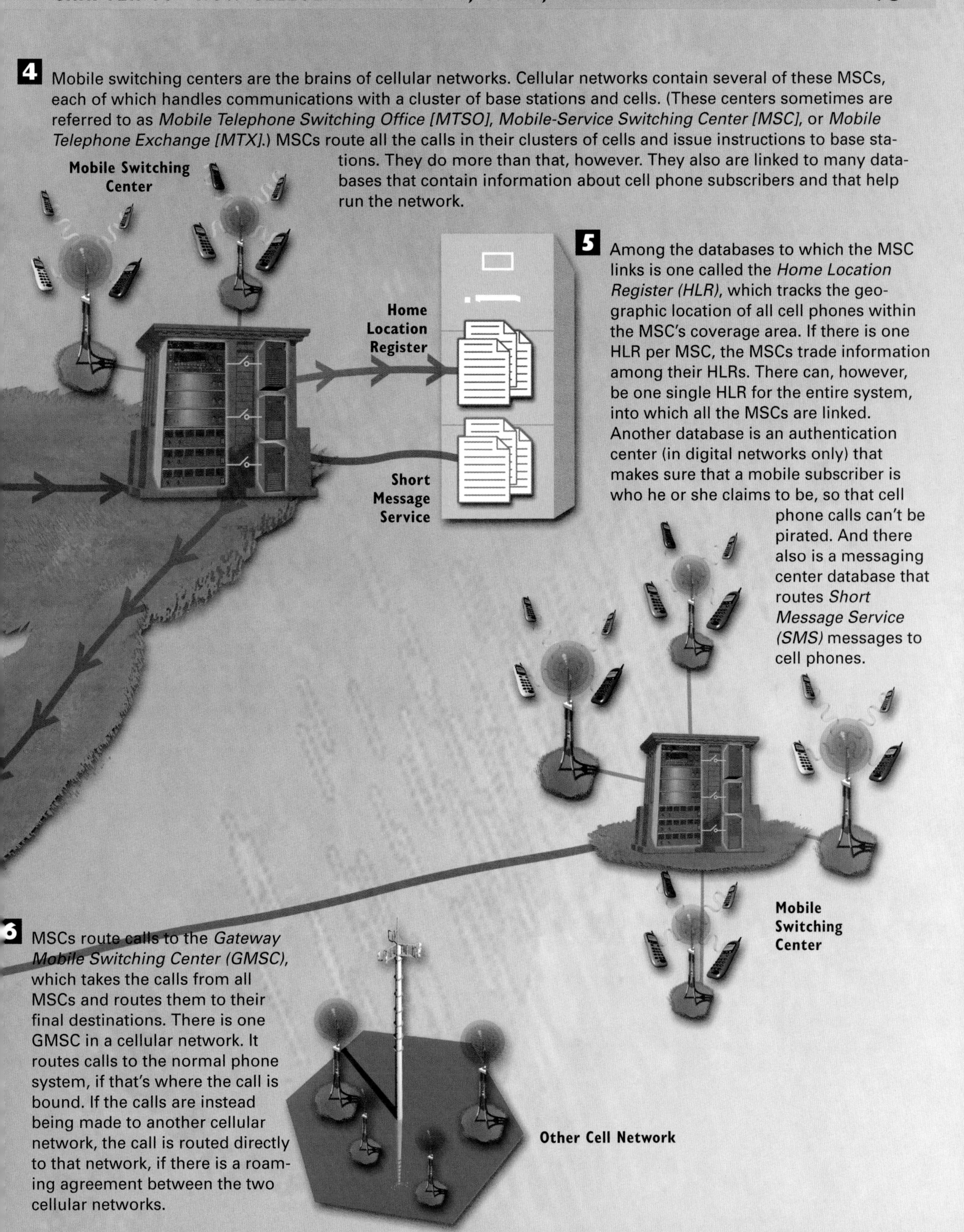

4 Mobile switching centers are the brains of cellular networks. Cellular networks contain several of these MSCs, each of which handles communications with a cluster of base stations and cells. (These centers sometimes are referred to as *Mobile Telephone Switching Office [MTSO]*, *Mobile-Service Switching Center [MSC]*, or *Mobile Telephone Exchange [MTX]*.) MSCs route all the calls in their clusters of cells and issue instructions to base stations. They do more than that, however. They also are linked to many databases that contain information about cell phone subscribers and that help run the network.

5 Among the databases to which the MSC links is one called the *Home Location Register (HLR)*, which tracks the geographic location of all cell phones within the MSC's coverage area. If there is one HLR per MSC, the MSCs trade information among their HLRs. There can, however, be one single HLR for the entire system, into which all the MSCs are linked. Another database is an authentication center (in digital networks only) that makes sure that a mobile subscriber is who he or she claims to be, so that cell phone calls can't be pirated. And there also is a messaging center database that routes *Short Message Service (SMS)* messages to cell phones.

6 MSCs route calls to the *Gateway Mobile Switching Center (GMSC)*, which takes the calls from all MSCs and routes them to their final destinations. There is one GMSC in a cellular network. It routes calls to the normal phone system, if that's where the call is bound. If the calls are instead being made to another cellular network, the call is routed directly to that network, if there is a roaming agreement between the two cellular networks.

How Cell Phones Connect to the Network

1. Cell phones have internal memory referred to as the *Number Assignment Module (NAM)*. Programmed into the NAM is the *Mobile Identification Number (MIN)*, which contains the wireless phone number; a number identifying the cell phone system with which it works, called the System ID, or SID; and information such as the features for which the customer has paid. The phone also contains an *Electronic Serial Number (ESN)*, which identifies the phone and helps guard against cell phone fraud.

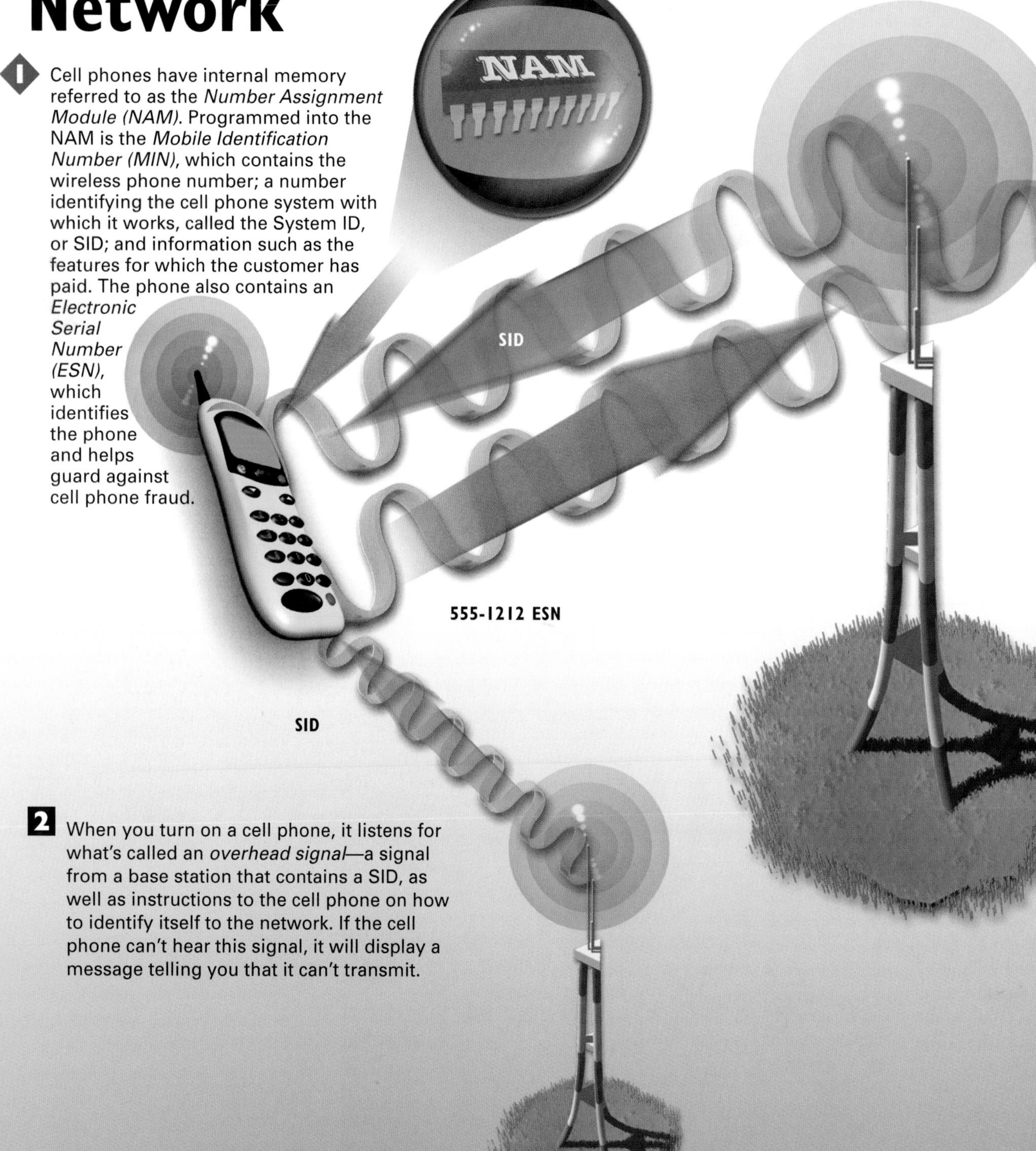

2. When you turn on a cell phone, it listens for what's called an *overhead signal*—a signal from a base station that contains a SID, as well as instructions to the cell phone on how to identify itself to the network. If the cell phone can't hear this signal, it will display a message telling you that it can't transmit.

3 Depending on its location, the phone might get signals from more than one base station. If that is the case, it detects which is the strongest signal, and then tunes to that base station. It repeats this every few minutes, so you're assured of a good connection when you walk or drive with your cell phone.

4 The phone compares the SID in the overhead signal to its own SID. If the two match, it means that the phone is in its home network. If they don't match, the phone will go into roaming mode, which allows it to connect to this non-home network. The subscriber, though, will be charged higher rates. In either event, the phone identifies itself to the network by sending its wireless phone number and its ESN.

5 This information is sent from the base station to the *Mobile Switching Center (MSC)*.

6 The MSC stores this information in the *Home Location Register (HLR)* database, so it now knows the precise location of the cell phone and with which base station it is communicating. The MSC uses this information to route calls to the cell phone and to better manage loads on the system. Every several minutes, when your phone is turned on, it exchanges information with the base station, and that information is relayed to the MSC and stored in the HLR. In this way, the system knows where you are, even if you're not making or receiving a call.

Wireless Tidbits

Base stations continually monitor the strength of your phone's signal and, depending on its strength, order the phone to increase or decrease its power output. (This happens every few seconds on an AMPS network, or many times a second in a CDMA network.) The base station tries to strike a balance between maintaining a clear connection and not making the power output too strong so that it interferes with other phones on the network.

How Cell Phones Make Calls

1 Cell phones use *pre-origination dialing*, which means that you dial the number you're planning to call before you actually connect to the network. When you dial a cell phone, you hear a series of dial tones. Those dial tones serve no functional purpose because cellular networks don't use dial tones. The dial tones are there instead so that you get some kind of auditory feedback to let you know that you're dialing the phone.

2 After you dial your call and press the SEND key, the network makes sure that your phone is tuned to the strongest wireless channel for making a phone call. The MSC makes this determination and sends that information to the phone through the base station.

3 After the base station ensures that you're tuned to the best channel, your cell phone transmits its MIN, its ESN, and the number that you dialed.

6 The GMSC and the normal telephone system or other cellular network exchange signaling tones, which are used to make sure that the connection between the systems are working properly. When it's clear that they are, you'll hear the normal ringing sound to tell you that your call has been made. When a person picks up the phone, you talk to them as you would on a normal phone—your call is complete.

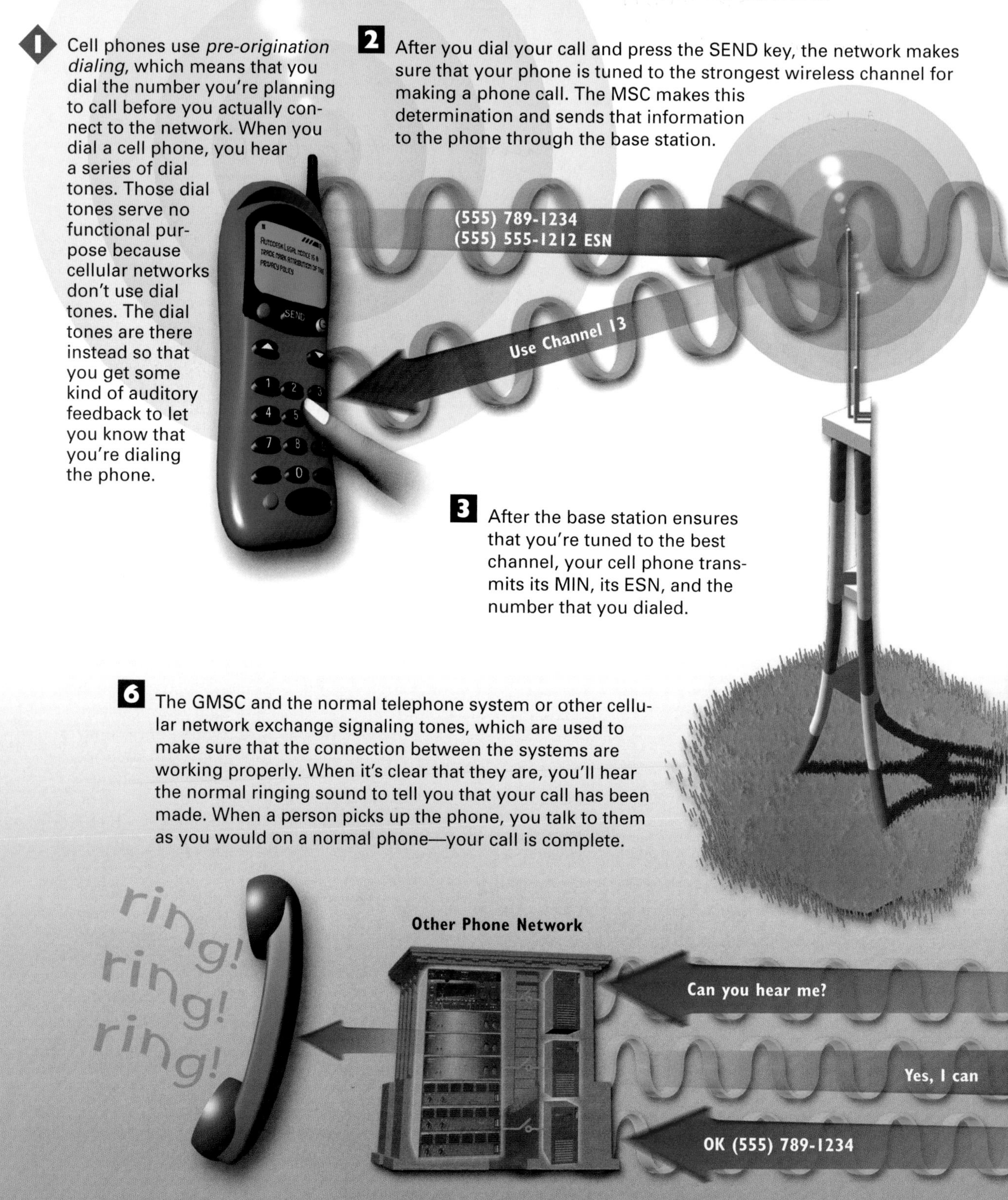
SEND
(555) 789-1234
(555) 555-1212 ESN
Use Channel 13
ring!
ring!
ring!
Other Phone Network
Can you hear me?
Yes, I can
OK (555) 789-1234

4 The information is relayed by the base station to the *Mobile Switching Center (MSC)*. The MSC verifies the authenticity of the cell phone, and then routes the call request to the *Gateway Mobile Switching Center (GMSC)*.

5 The GMSC routes the call to the normal telephone system (or to another cellular network if it's a cell phone-to-cell phone call).

How Cell Phones Receive Calls

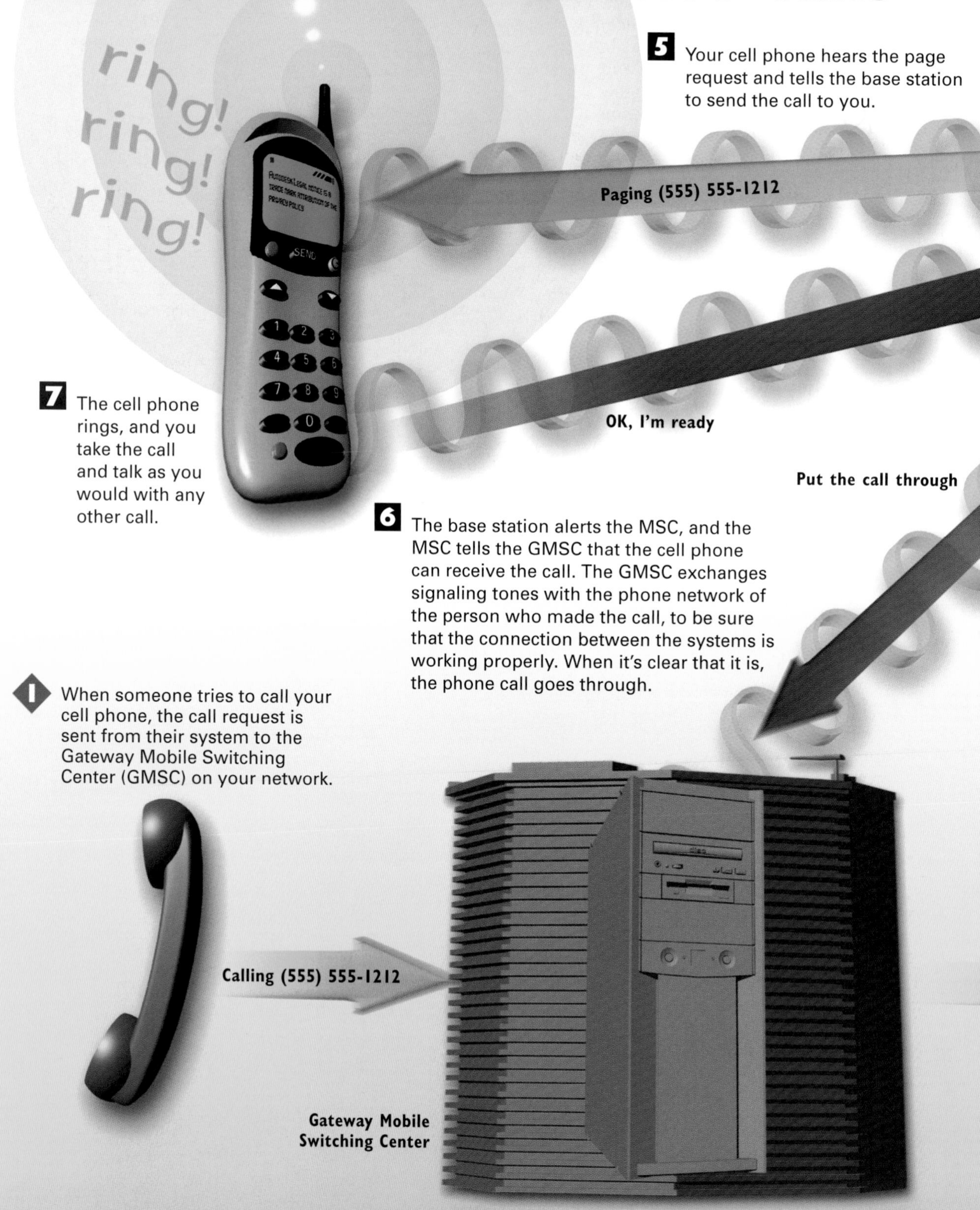
ring!
ring!
ring!
Paging (555) 555-1212
OK, I'm ready
Put the call through
Calling (555) 555-1212
Gateway Mobile
Switching Center

1 When someone tries to call your cell phone, the call request is sent from their system to the Gateway Mobile Switching Center (GMSC) on your network.

5 Your cell phone hears the page request and tells the base station to send the call to you.

6 The base station alerts the MSC, and the MSC tells the GMSC that the cell phone can receive the call. The GMSC exchanges signaling tones with the phone network of the person who made the call, to be sure that the connection between the systems is working properly. When it's clear that it is, the phone call goes through.

7 The cell phone rings, and you take the call and talk as you would with any other call.

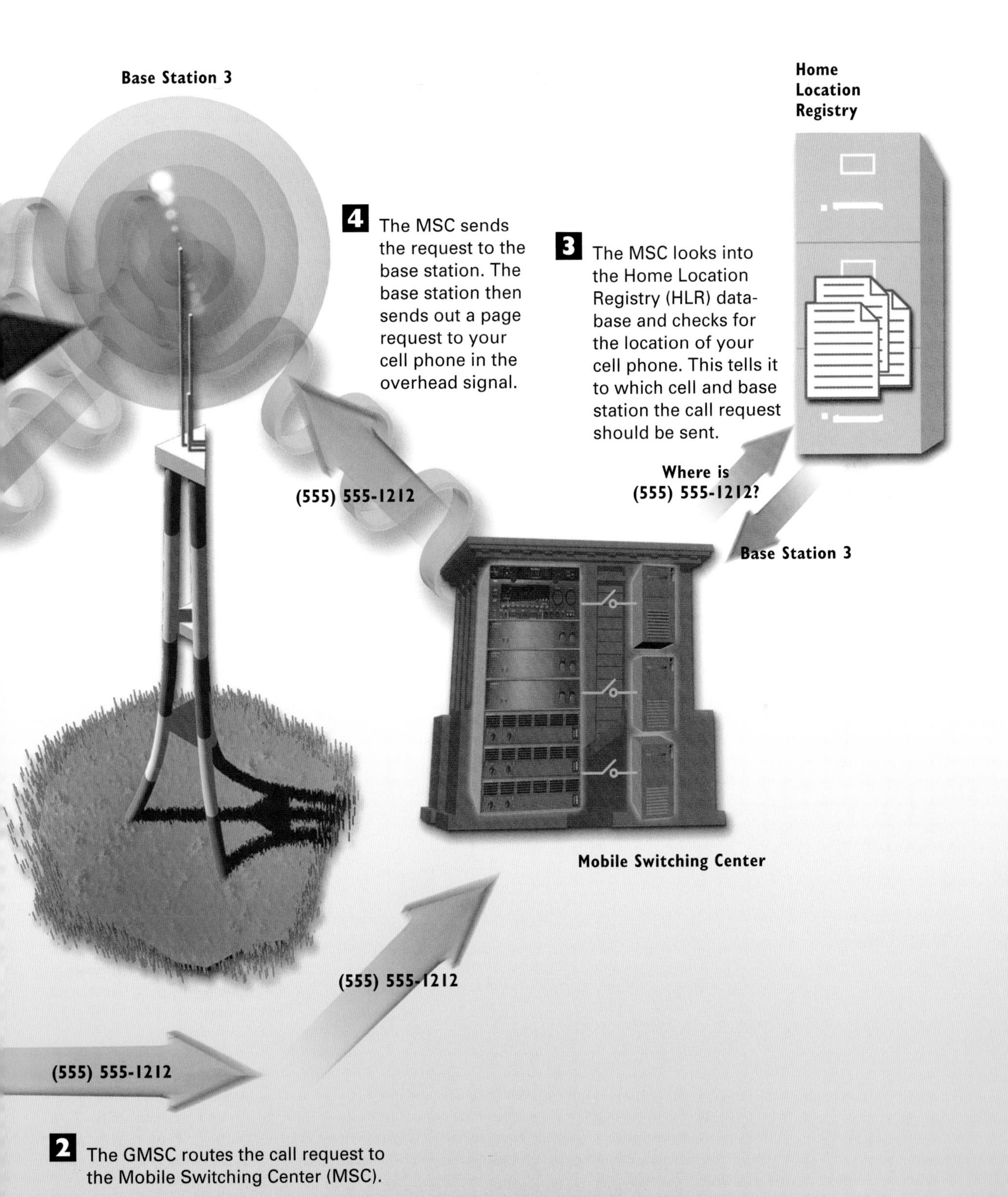

4 The MSC sends the request to the base station. The base station then sends out a page request to your cell phone in the overhead signal.

3 The MSC looks into the Home Location Registry (HLR) database and checks for the location of your cell phone. This tells it to which cell and base station the call request should be sent.

2 The GMSC routes the call request to the Mobile Switching Center (MSC).

How Cells and Handoffs Work

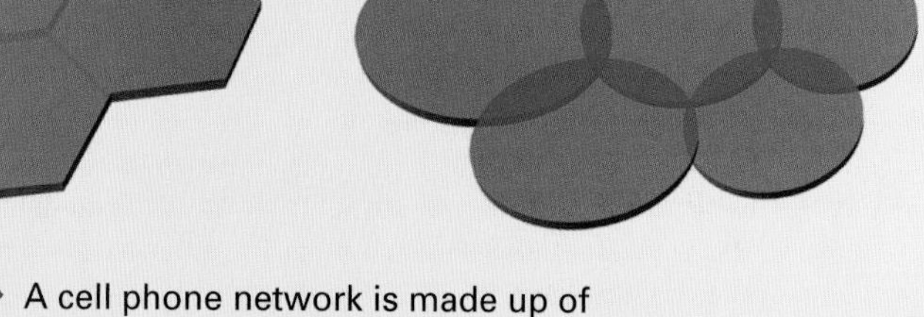

1 A cell phone network is made up of many overlapping cells, each of which has a base station in it to serve cell phones in its own cell. Although, on paper, cells generally are drawn as they are here—as hexagons—in real life they are approximately overlapping circles, as you can see in the adjacent illustration.

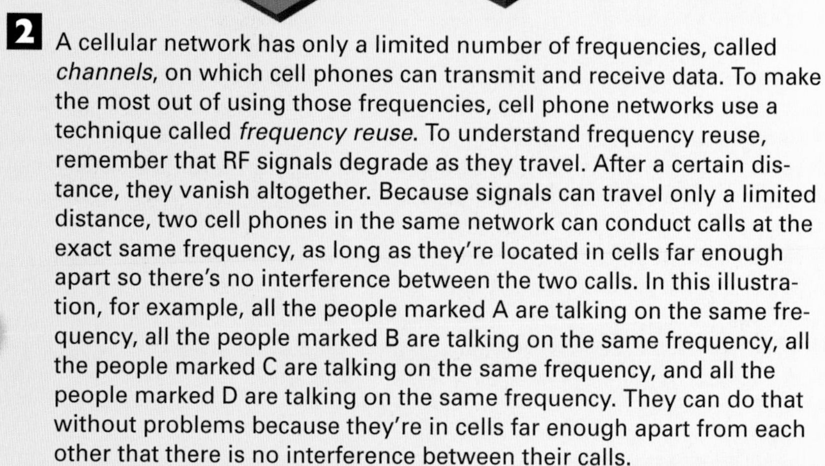

2 A cellular network has only a limited number of frequencies, called *channels*, on which cell phones can transmit and receive data. To make the most out of using those frequencies, cell phone networks use a technique called *frequency reuse.* To understand frequency reuse, remember that RF signals degrade as they travel. After a certain distance, they vanish altogether. Because signals can travel only a limited distance, two cell phones in the same network can conduct calls at the exact same frequency, as long as they're located in cells far enough apart so there's no interference between the two calls. In this illustration, for example, all the people marked A are talking on the same frequency, all the people marked B are talking on the same frequency, all the people marked C are talking on the same frequency, and all the people marked D are talking on the same frequency. They can do that without problems because they're in cells far enough apart from each other that there is no interference between their calls.

4 When someone with a cell phone travels from one cell to another, there needs to be a *handoff* of the call between cells because the base station in the first cell no longer will be able to contact the cell phone when it travels to the second cell. In a handoff, the responsibility for communicating with the cell phone switches from the base station in one cell to the base station of another cell.

Wireless Tidbit

Cell phone networks have "holes" in them—areas in which calls can't be sent or received. Wireless phone signals are sent along "lines of sight," which means that hilly terrain can interrupt signals and create holes. That's why, when you're driving in a car and talking, you might come into spots where you can't talk on your cell phone. Because of this, cell phone networks are powerfully affected by nature. For example, when the leaves fall off the trees in the autumn, engineers might have to retune their cell networks to take less interference into account. When the leaves grow back again in the spring, they again have to retune their networks.

3 Many different things determine the size of cells in a cell phone network. However, one basic thing that determines the maximum cell size is the frequency at which the network is licensed to operate. Different types of cellular phone networks are licensed to operate at different frequencies. For example, many cellular networks are licensed to operate in the 800 MHz frequency, whereas PCS networks are licensed to operate in the 1900 MHz frequency. An 800 MHz signal can travel much farther than a 1900 MHz signal at the same power, so cells in 800 MHz networks can be much larger than cells in 1900 MHz networks. Other factors that determine cell size are antenna height and transmitter power, among many others.

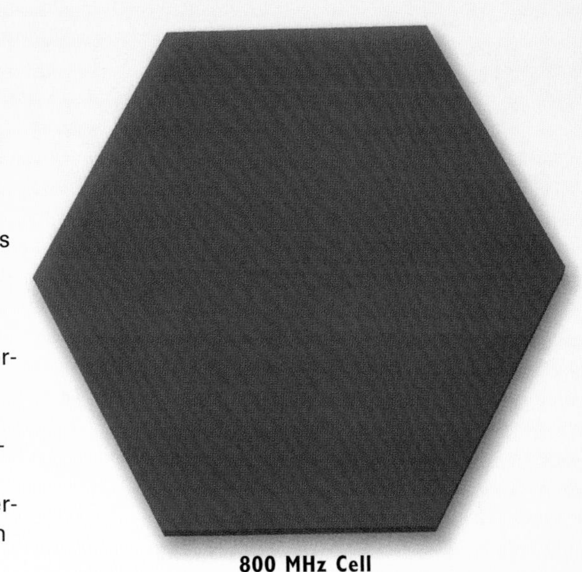

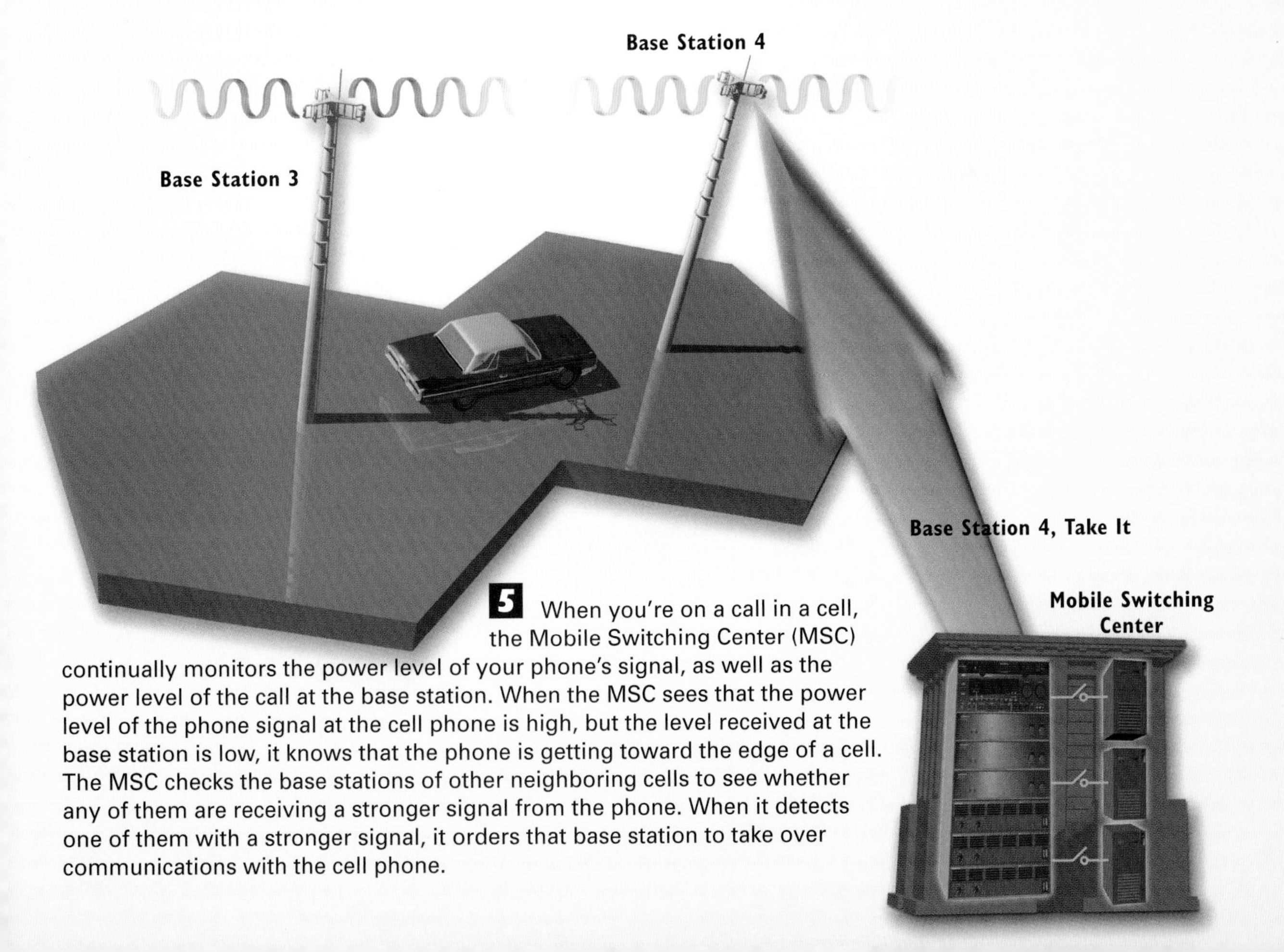

5 When you're on a call in a cell, the Mobile Switching Center (MSC) continually monitors the power level of your phone's signal, as well as the power level of the call at the base station. When the MSC sees that the power level of the phone signal at the cell phone is high, but the level received at the base station is low, it knows that the phone is getting toward the edge of a cell. The MSC checks the base stations of other neighboring cells to see whether any of them are receiving a stronger signal from the phone. When it detects one of them with a stronger signal, it orders that base station to take over communications with the cell phone.

CHAPTER

11

How Cellular Telephones Work

AT the heart of the wireless communications revolutions is a device so small and slim it can fit in the palm of your hand, and yet contains such sophisticated electronics, computing, and communications technologies that they were inconceivable decades ago. We're talking, of course, about the cell phone, the device that has become ubiquitous: Everywhere you go—from the grocery store to city streets, cafés, and, of course, automobiles—it seems to be glued to people's ears.

The devices themselves contain an astonishing array of technologies taken from different industries. There are liquid crystal displays, microprocessors, antennas, amplifiers, circuit boards, microphones, speakers, digital signal processors, and much more. If you take a cell phone apart, you'd be amazed at all the electronics squashed into such a small space.

Of course, you don't want to take one apart, so the first illustration in this chapter shows you what the inside of a cell phone looks like, and explains what all the important components do.

The cell phone by itself, of course, can't do anything—for that, it needs a network. In the previous chapter, you learned how cell phone networks work. In this chapter, you'll look at various underlying technologies that work closely with the phones. You'll see, for example, how control and communications channels work in concert with cell phones to send and receive data.

If you've tried to buy a cell phone and cell phone service, you know what a mind-boggling number of acronyms and incompatible technologies you have to wend your way through when making a decision. So, the rest of the illustrations in this chapter will help you understand the most important of those technologies.

Phones can use either digital or analog technology. Analog phones are older, and eventually they probably no longer will be manufactured as the world goes digital. Digital phones offer far more services, but to deliver them, they use more complex technology.

The acronyms you'll most likely encounter in the cell phone world are *PCS* (Personal Communication Service), *TDMA* (Time Division Multiple Access), *CDMA* (Code Division Multiple Access), and GSM (Global System for Mobile Communications). PCS doesn't really refer to a single, discrete technology; instead, it's a more general term describing digital cellular systems that offer a wide range of services, such as text messaging (known as Short Message Service, or SMS) and access to the Web and e-mail. GSM, primarily used in Europe, also doesn't describe a single technology, but rather a host of related technologies. PCS, in fact, is based on GSM, and is essentially the U.S. version of a GSM system.

TDMA and CDMA, on the other hand, are specific technologies. They're digital technologies that use different methods of allowing many different people to share the same frequency for cell phone use.

A Cutaway View of a Cellular Telephone

Speaker This changes the signals inside the phone into sounds that you can understand.

Liquid Crystal Display (LCD) or plasma display The display does more than just display the number you're dialing—it can display SMS text messages, menus, Web pages, and e-mail. Because of that, cell phone displays are getting larger and offering better resolutions.

Wireless Tidbit

The ring tones of cell phones are familiar not just to people, but to animals as well. In fact, they're so common that birds have started copying the tones in their bird calls. Several different types of birds have been heard to copy the tones. Ornithologist Helene Lampe says that the bird that most commonly copies the ring tones are starlings.

Keyboard You can type in messages as well as phone numbers. The part of the keyboard you don't see is a circuit board that translates the key you press into commands that the phone can understand.

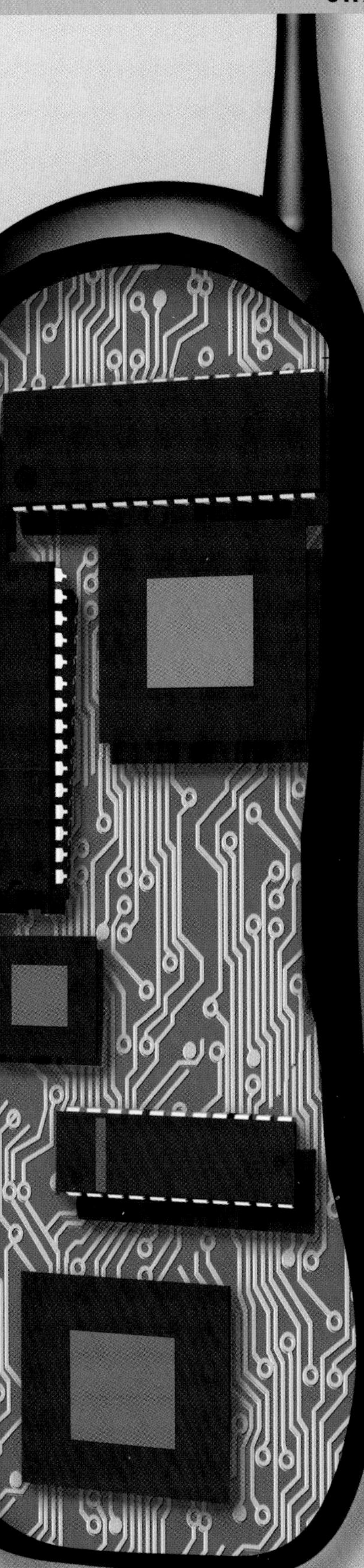

Antenna This receives and sends RF signals.

Analog-to-Digital and Digital-to-Analog chips The analog-to-digital chip takes analog signals from the microphone and translates them to digital signals that need to be processed in the phone before they're sent. The digital-to-analog chip takes the received digital signals in the phone and translates them to analog waves that can be played through the speakers.

Digital Signal Processor (DSP) Handles compressing and decompressing the digital signal sent and received by the phone. Signals are compressed to save bandwidth space during transmission, and the DSP compresses the signal before it's sent and decompresses signals when they're received. It also does modulation and demodulation and performs error correction.

ROM and flash memory chips Contain storage for the telephone's operating system and for information such as your telephone directory of contacts.

Battery This supplies the electricity for your phone

Circuit board This contains all the electronics of the phone.

Amplifiers Amplify the signals received from the antenna and the signals that are to be sent by the antenna.

Microprocessor This is command central for your phone. It's the brains of the phone. It handles shuttling information from and to the keyboard and display, coordinates the work of all the electronics and chips on the circuit board, and contains all the logic and intelligence in the phone.

Microphone This takes your voice and changes it into analog electrical signals.

How Cellular Channels Work

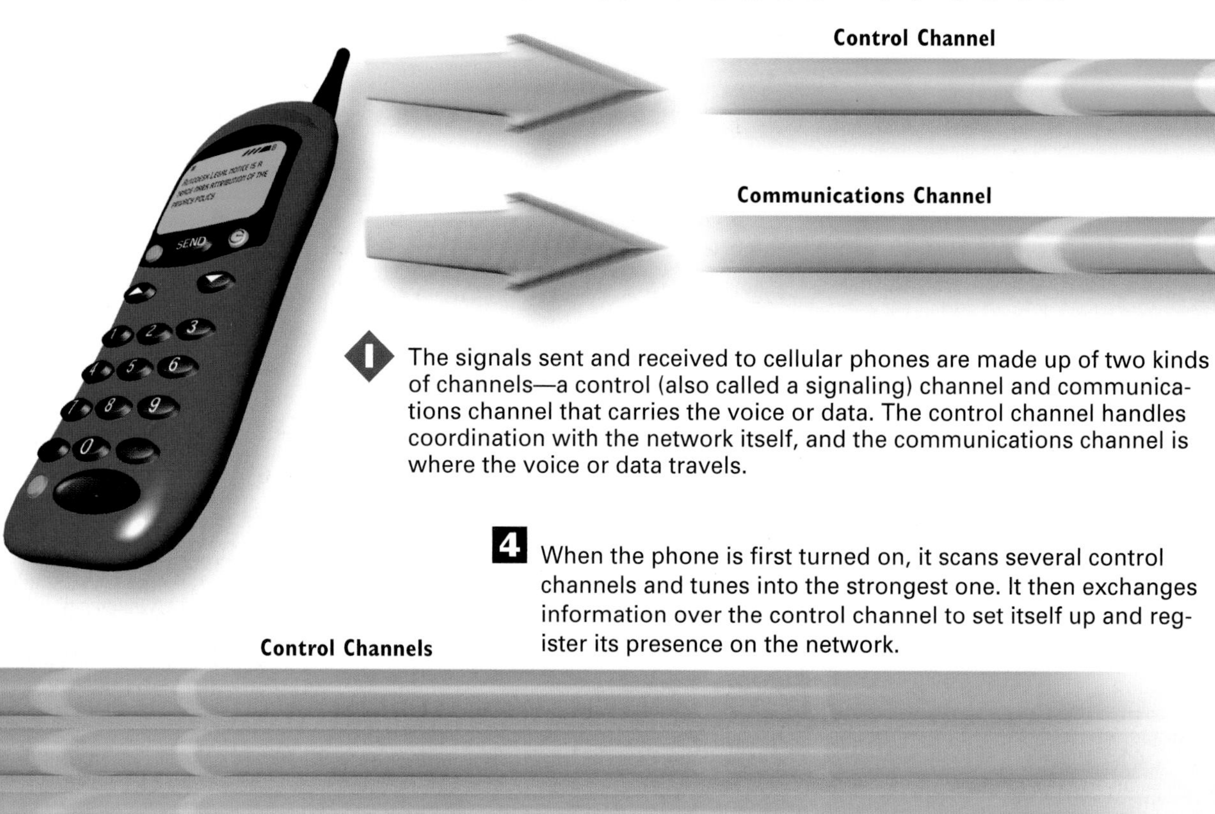

1 The signals sent and received to cellular phones are made up of two kinds of channels—a control (also called a signaling) channel and communications channel that carries the voice or data. The control channel handles coordination with the network itself, and the communications channel is where the voice or data travels.

4 When the phone is first turned on, it scans several control channels and tunes into the strongest one. It then exchanges information over the control channel to set itself up and register its presence on the network.

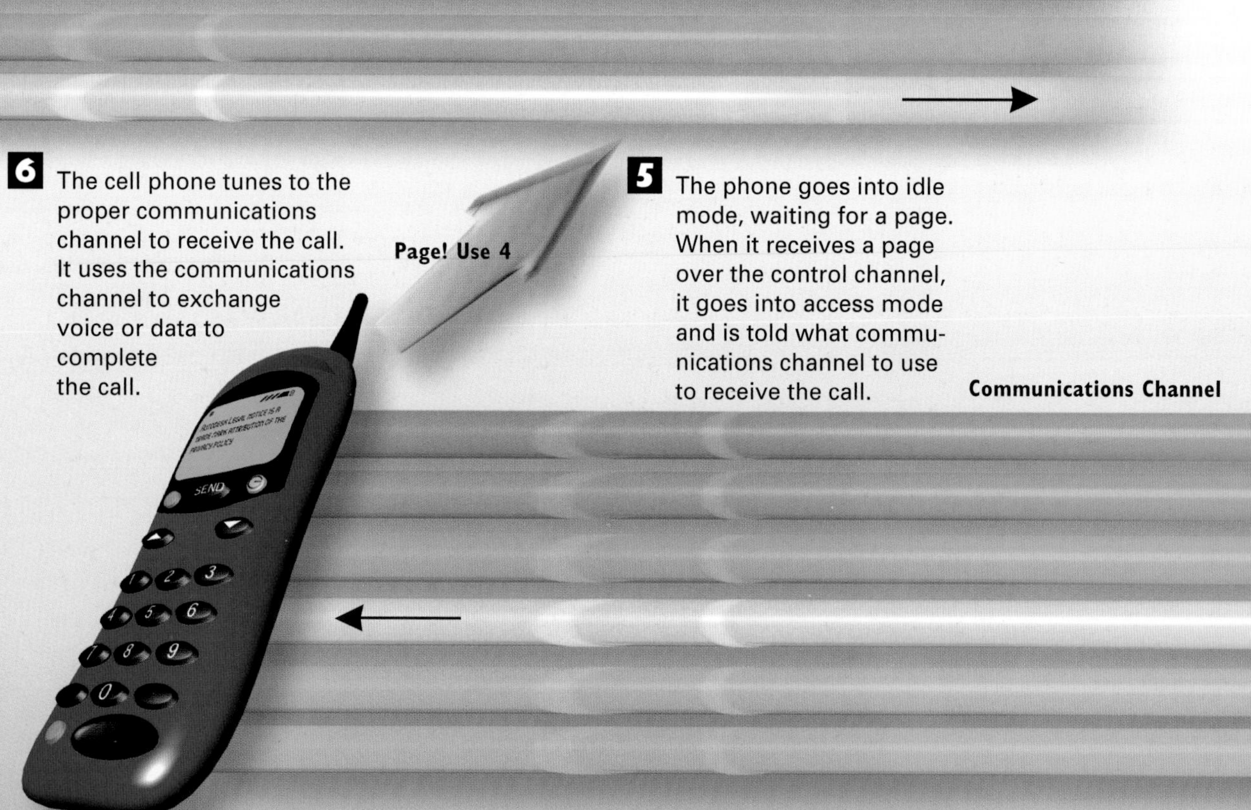

5 The phone goes into idle mode, waiting for a page. When it receives a page over the control channel, it goes into access mode and is told what communications channel to use to receive the call.

6 The cell phone tunes to the proper communications channel to receive the call. It uses the communications channel to exchange voice or data to complete the call.

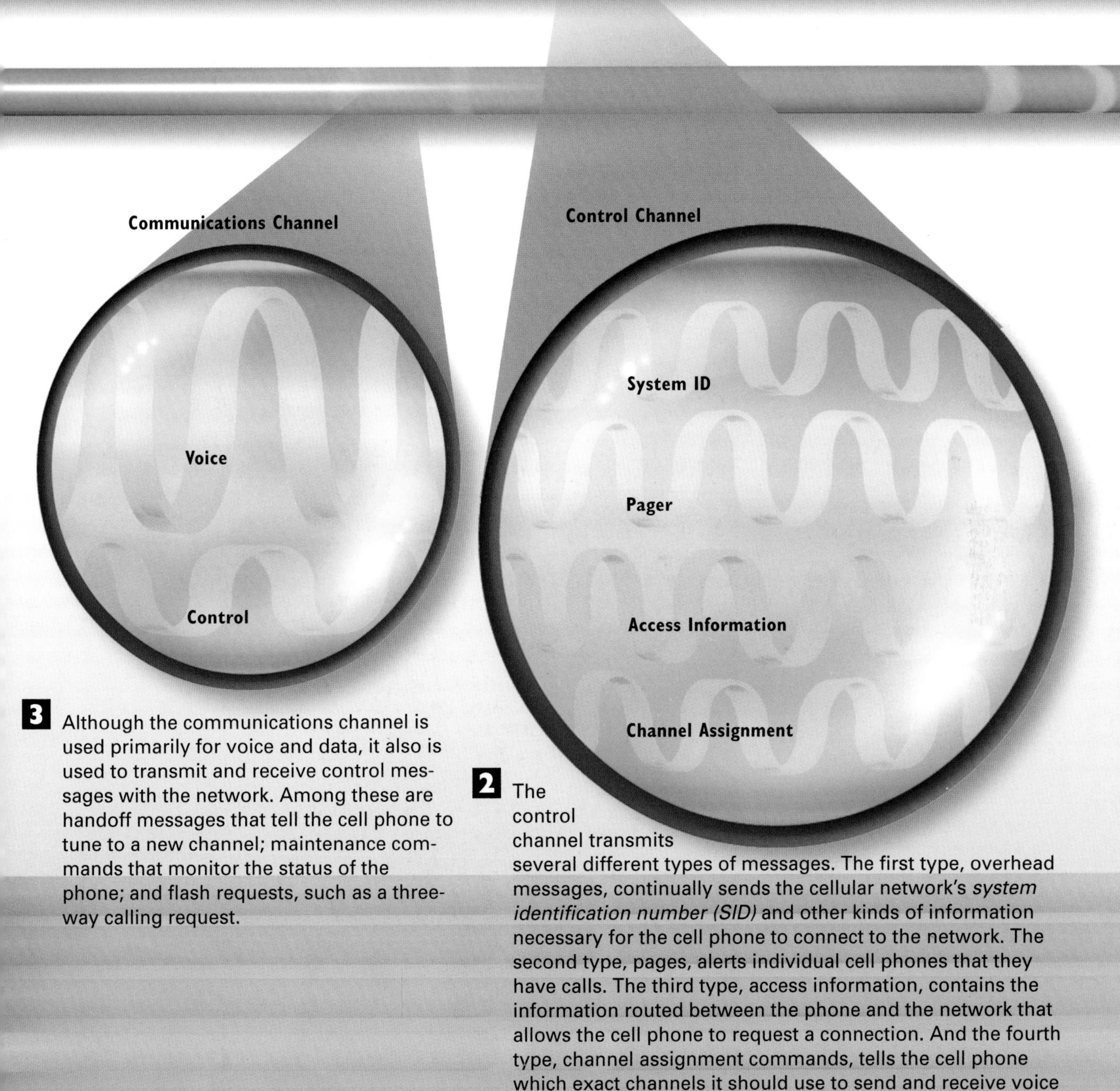

3 Although the communications channel is used primarily for voice and data, it also is used to transmit and receive control messages with the network. Among these are handoff messages that tell the cell phone to tune to a new channel; maintenance commands that monitor the status of the phone; and flash requests, such as a three-way calling request.

2 The control channel transmits several different types of messages. The first type, overhead messages, continually sends the cellular network's *system identification number (SID)* and other kinds of information necessary for the cell phone to connect to the network. The second type, pages, alerts individual cell phones that they have calls. The third type, access information, contains the information routed between the phone and the network that allows the cell phone to request a connection. And the fourth type, channel assignment commands, tells the cell phone which exact channels it should use to send and receive voice and data.

Understanding the Difference Betwee Analog and Digital Systems

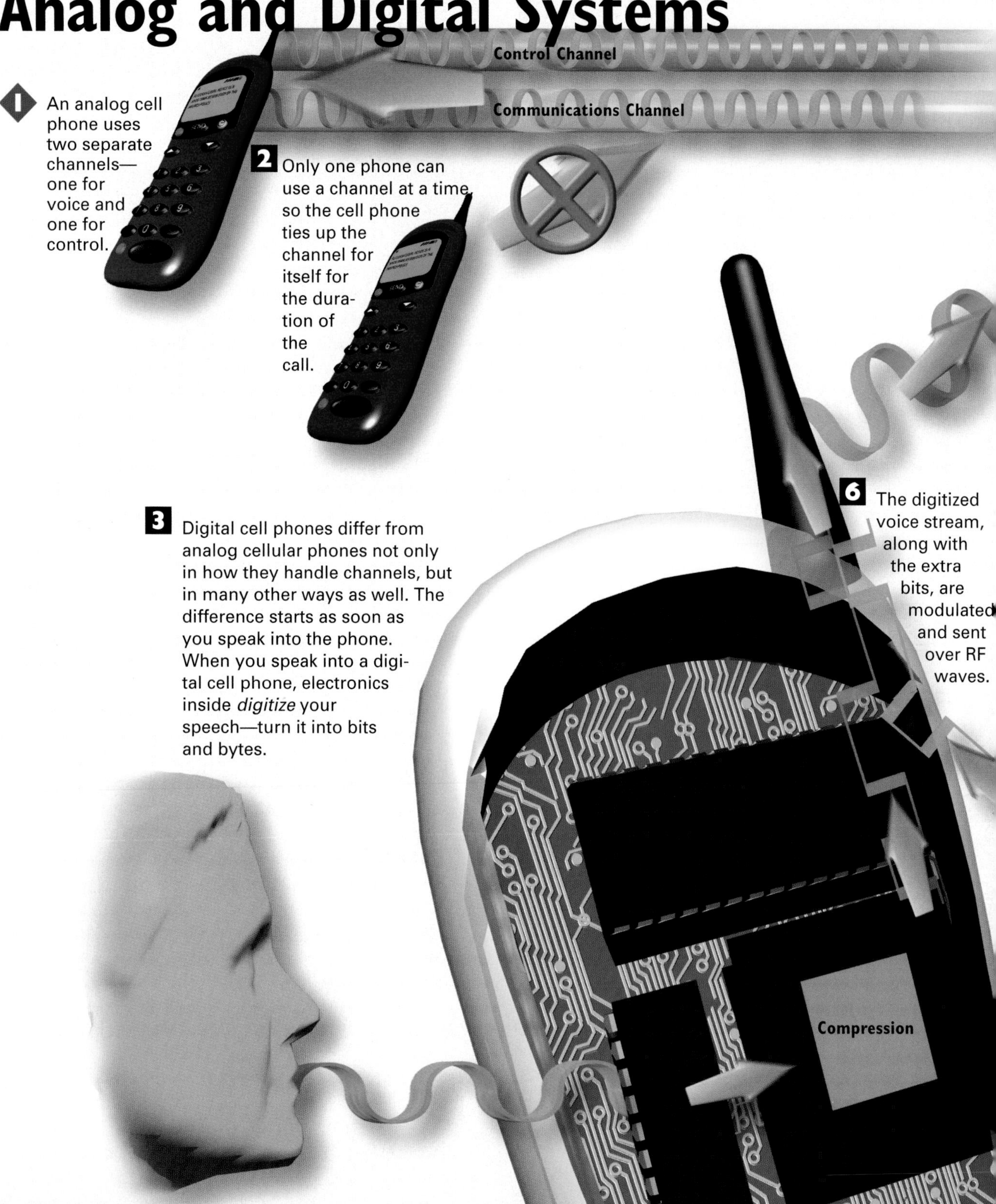

1 An analog cell phone uses two separate channels—one for voice and one for control.

2 Only one phone can use a channel at a time, so the cell phone ties up the channel for itself for the duration of the call.

3 Digital cell phones differ from analog cellular phones not only in how they handle channels, but in many other ways as well. The difference starts as soon as you speak into the phone. When you speak into a digital cell phone, electronics inside *digitize* your speech—turn it into bits and bytes.

6 The digitized voice stream, along with the extra bits, are modulated and sent over RF waves.

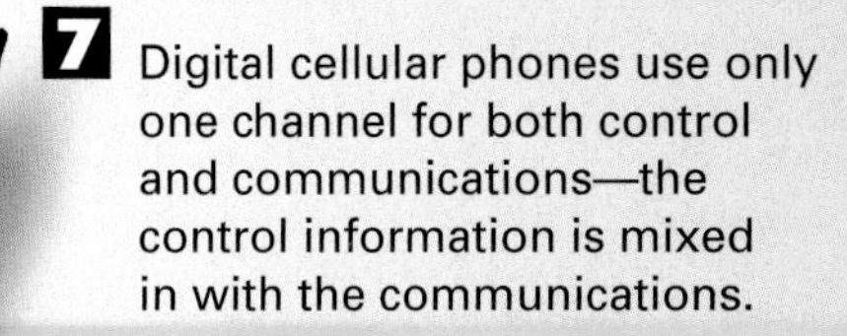

7 Digital cellular phones use only one channel for both control and communications—the control information is mixed in with the communications.

8 Many digital cell phones can simultaneously use the same channel for control and communications.

Wireless Tidbit

Uncompressed digitized voice produces a stream of data at a rate of 64 kilobits per second, which means that it would take up more than 64 kHz of bandwidth per call, which is quite wasteful, and expensive for a cellular network operator. That's a large reason why compression is used. Depending on the cell phone system used, the amount of digital voice compression varies, ranging from about 5 to 1, all the way up to 64 to 1.

Base Station Controller

9 When the digital stream reaches the base station, electronics separate the voice channel from the control channel and route both through the rest of the cellular system.

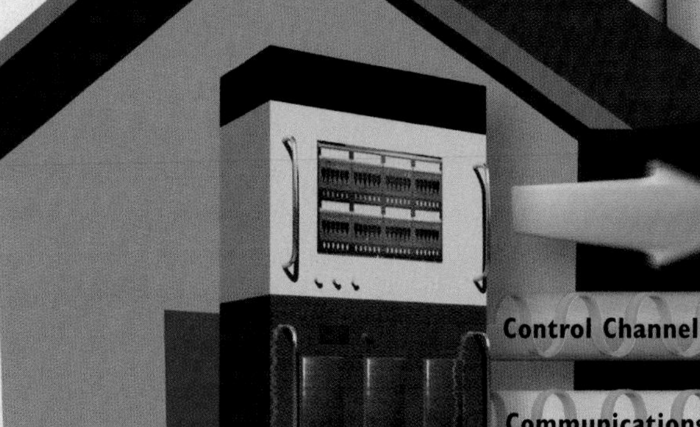

5 Extra bits are added to the digitized voice. Some of these bits are required for control information, and others are used as a way to correct any errors that might be introduced during transmission.

4 The electronics also compress your speech using speech compression technology—often called voice coding. This reduces the size of the signal. The more the voice is compressed, the lower the quality of the received sound.

Understanding PCS Systems

800 MHz Cellular

TDMA

CDMA

1900 MHz PCS

GSM

1 The term *Personal Communications Service (PCS)* doesn't refer to a single technology, but rather it's a general name for newer cellular systems that offer many kinds of cellular communications services. PCS systems are licensed to use the 1900 MHz FR spectrum band. Earlier cellular systems, such as analog systems, typically use the 800 MHz band.

2 PCS systems are all-digital systems. There is no single standard for them, though: They can use a variety of standards and technologies, including *Time Division Multiple Access (TDMA)*, *Code Division Multiple Access (CDMA)*, and the *Global Systems for Mobile Communications (GSM)*. (To understand how those technologies work, turn to the illustration on the following pages.)

4 The *Federal Communications Commission (FCC)* limits the power levels of cellular transmitters. Because of this, PCS networks must have towers—and cells—closer together than 800 MHz cellular networks.

800 MHz System

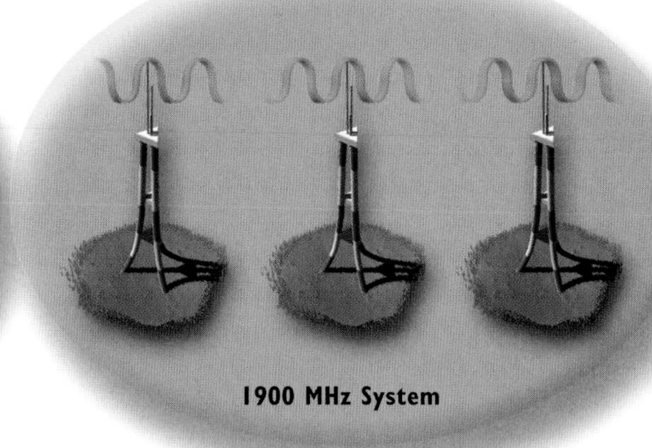

1900 MHz System

5 PCS networks offer many communications services that earlier cellular technologies didn't have. For example, they offer the *Short Message Service (SMS)* that allows for text messaging between cell phones, as well as Internet access and other features.

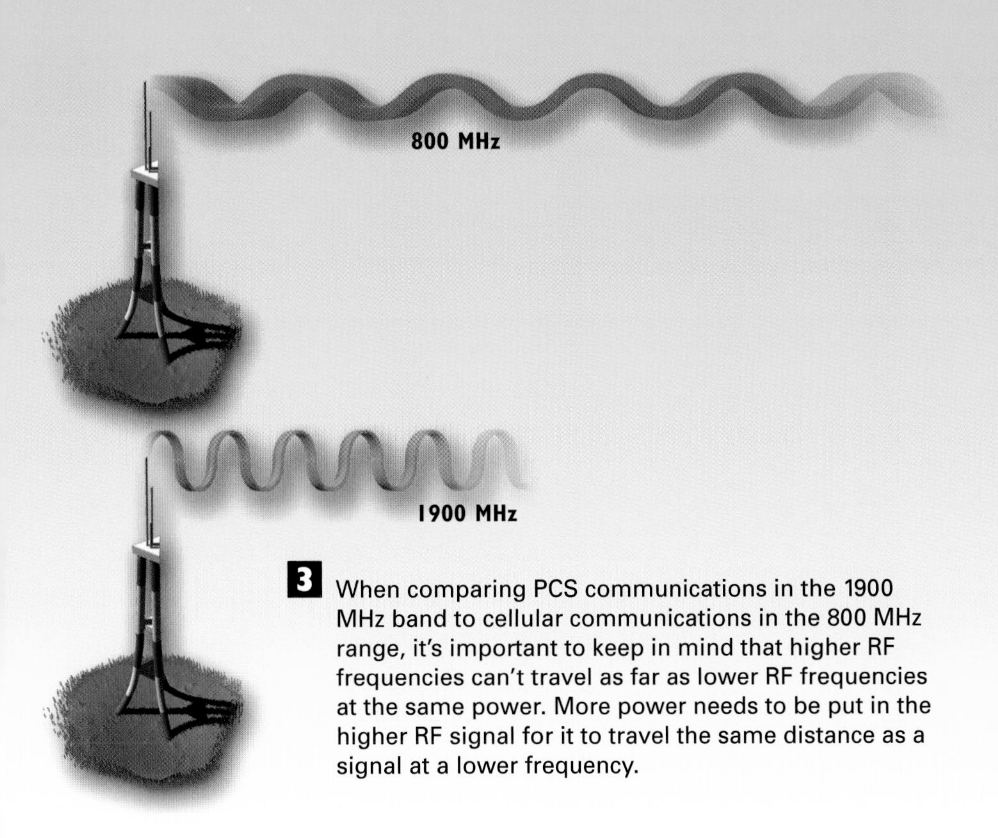

3 When comparing PCS communications in the 1900 MHz band to cellular communications in the 800 MHz range, it's important to keep in mind that higher RF frequencies can't travel as far as lower RF frequencies at the same power. More power needs to be put in the higher RF signal for it to travel the same distance as a signal at a lower frequency.

How TDMA, CDMA, and GSM Work

1 *Time Division Multiple Access (TDMA)* is a digital technique that allows several cell phones to use the same channel simultaneously. To do this, it divides the channel into sequential time slots.

6 1 2 3 4 5 6 1 2

2 Each cell phone is assigned its own specific time slot during its call. It sends and receives bursts of data in that time slot. The data is sent so quickly that, even though each cell phone has a time devoted to it on the channel, all the communications appear to happen simultaneously. When the call ends and the phone is in another call, it is assigned another slot.

3 *Code Division Multiple Access (CDMA)* is another digital technique that allows several cell phones to share the same channel simultaneously. When a cell phone gets onto a CDMA system, it is assigned to what is called a coded channel.

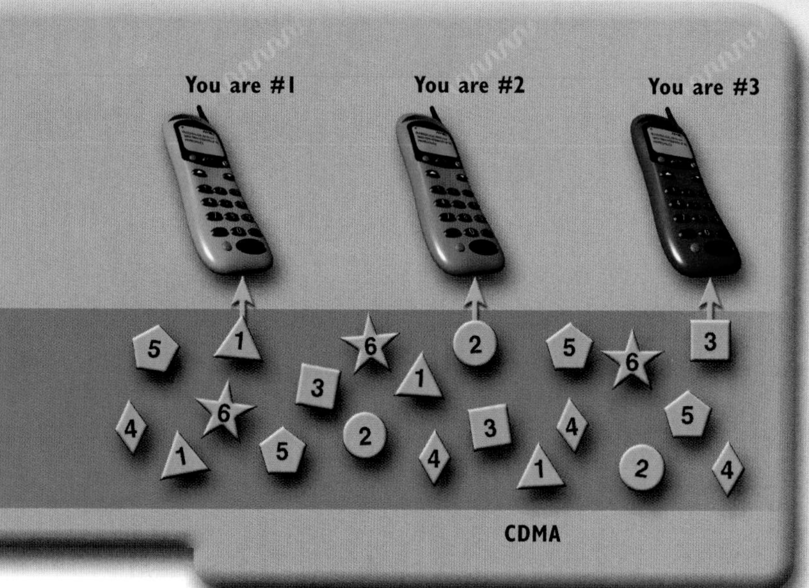

4 All the digital voice, data, and control information for many different phones is sent simultaneously along a wideband channel that contains information from many coded channels inside it.

5 Cell phones are able to get information just from their coded channel from the wideband channel, and send coded information through the wideband channel as well.

6 The Global System for Mobile Communications (GSM) is a standard for digital cellular communications developed in Europe and put into effect in 1992. It is designed so that Europe can operate with a single cellular standard. It uses TDMA as its way of communicating, and operates in different frequencies in different countries: at 450 MHz, 900 MHz, 1800 MHz, and 1900 MHz. The U.S. PCS system is based on GSM and operates at 1900 MHz.

Wireless Tidbit

The GSM standard and its variants, including PCS in the United States, has become the most popular cell phone standard in the world. As of the middle of 2001, a half a billion people were using some variant of GSM, according to cell phone maker Ericsson.

7 GSM subscribers use a subscriber identity module (SIM) card, which identifies the cell phone user and contains on it the mobile phone number; the mobile electronic identity number (MEIN), which is a kind of serial number; and other identifying information.

8 The SIM card can be taken out of the phone and carried around. So, when someone from one country goes to another country, all they need to do is put their own SIM card in the cell phone, and they'll be able to make phone calls. The calls will be billed to their SIM card.

How Short Message Service (SMS) Works

Mobile Switching Center

I got the job!

1 SMS allows text messages to be sent to cell phones from other cell phones, over e-mail, from personal digital assistants (PDAs), and other similar devices. The message is typed out, and then sent to a specific mailbox, which eventually will be routed to a cell phone.

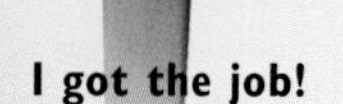

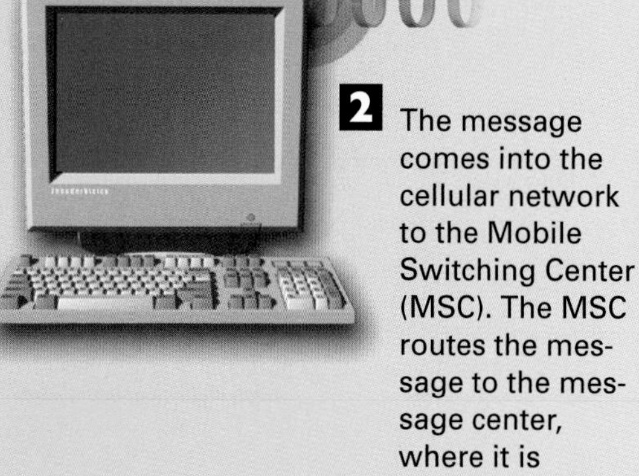

2 The message comes into the cellular network to the Mobile Switching Center (MSC). The MSC routes the message to the message center, where it is stored.

Messaging Center

3 The network checks for where the cellular phone is located, and then sends a signal to the base station of the nearest cell to alert the phone that an SMS message is on the way.

4 The base station sends out the alert. The alert is sent over the control channel.

5 The cell phone tunes to the channel where the message will be sent and receives the message. In most systems, SMS messages are sent on the control channel. (Note: Older cell phones might not be able to receive SMS messages. And with newer phones, some can receive messages but not send them.)

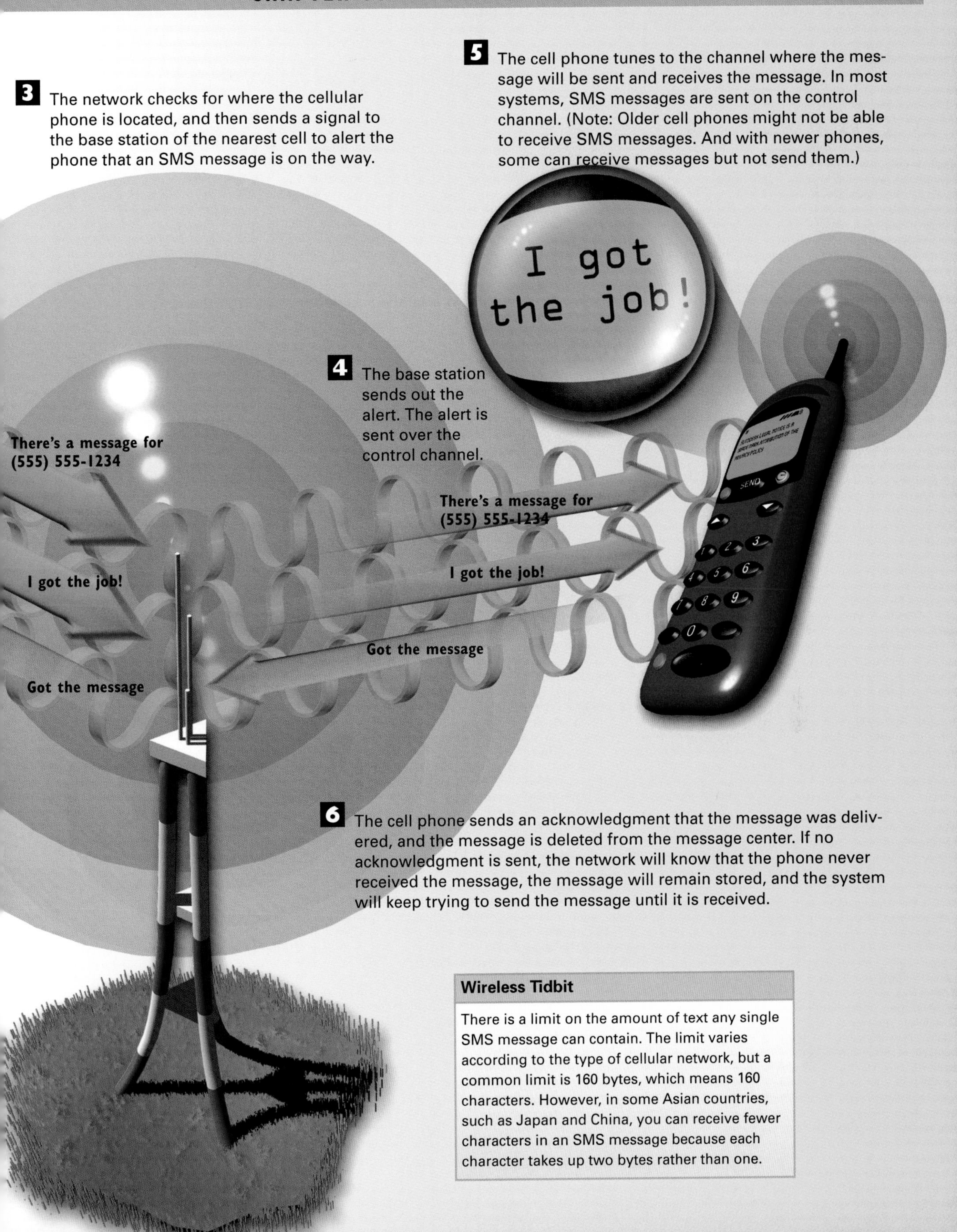
I got the job!
There's a message for (555) 555-1234
I got the job!
Got the message
There's a message for (555) 555-1234
I got the job!
Got the message
SEND

6 The cell phone sends an acknowledgment that the message was delivered, and the message is deleted from the message center. If no acknowledgment is sent, the network will know that the phone never received the message, the message will remain stored, and the system will keep trying to send the message until it is received.

Wireless Tidbit

There is a limit on the amount of text any single SMS message can contain. The limit varies according to the type of cellular network, but a common limit is 160 bytes, which means 160 characters. However, in some Asian countries, such as Japan and China, you can receive fewer characters in an SMS message because each character takes up two bytes rather than one.

CHAPTER

12

How Pagers Work

PAGERS, once worn primarily by doctors, plumbers, and others whose jobs required them to be reached in emergencies, have long ago reached the mainstream. Everyone from high-tech executives to teenagers now carry them on their belts so that they can be easily reached—and for some, it's become a sign of prestige, an announcement to the world that they're so important they need to be able to be contacted no matter the place or hour.

Pagers have become a huge part of our lives for several reasons. The devices are exceedingly small and can be unobtrusively carried. They're inexpensive, don't have high monthly fees, have a very long battery life, and generally work no matter where you are.

Pagers work like many other radio devices. They contain an antenna that receives the RF signal and electronics that translate that signal into letters or numbers and letters on an LCD screen. An individual pager is part of a network, which has an infrastructure similar to those of cell phones. Your pager has a unique code that identifies it, so it receives only pages that have been specifically sent to it. Unlike cell phones, pagers don't have transmitters (except in the case of two-way pagers).

The first pagers could do nothing more than alert you to the fact that someone wanted to get in touch with you, and show their phone number or some other numeric code. Since then, there have been significant advances in paging. Broadcast paging allows messages to be sent to many people, not just one. It can be used to deliver real-time news, stock quotes, and other information to people who want it.

Most notable has been the advent of two-way paging. Two-way pagers allow you not just to receive messages, but to send them as well. Becoming increasingly popular are two-way pagers that allow you to be alerted when you have e-mail, and then let you check your e-mail, and even to respond to it. Doing that requires you to work with your employer (if you want to get your business-related e-mail), or with your Internet service provider (ISP) if you want to get and send your personal e-mail.

Although two-way pagers—especially those that let you receive and send e-mail—are becoming increasingly popular, it's not yet clear how popular more traditional one-way pagers will remain. The newest cell phones have a pager-like function called Short Message Service (SMS), which allows cell phones to send and receive pager-like messages. For more information about cell phones and SMS, turn to Chapter 11, "How Cellular Telephones Work."

A Cutaway View of a Pager

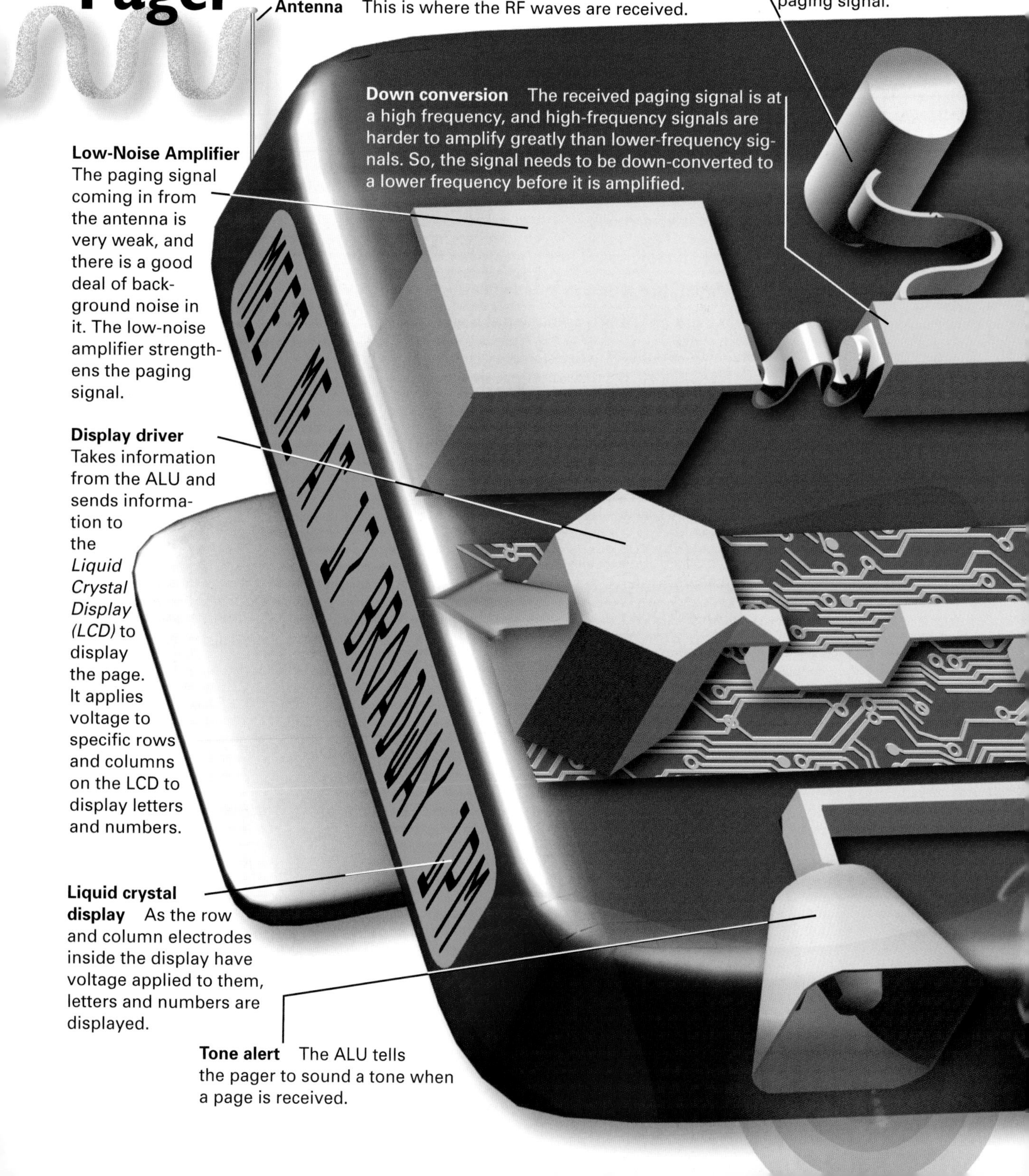

Antenna This is where the RF waves are received.

Oscillator An oscillator produces a wave that is used to down-convert the paging signal.

Down conversion The received paging signal is at a high frequency, and high-frequency signals are harder to amplify greatly than lower-frequency signals. So, the signal needs to be down-converted to a lower frequency before it is amplified.

Low-Noise Amplifier The paging signal coming in from the antenna is very weak, and there is a good deal of background noise in it. The low-noise amplifier strengthens the paging signal.

Display driver Takes information from the ALU and sends information to the *Liquid Crystal Display (LCD)* to display the page. It applies voltage to specific rows and columns on the LCD to display letters and numbers.

MEET ME AT 17 BROADWAY 1PM

Liquid crystal display As the row and column electrodes inside the display have voltage applied to them, letters and numbers are displayed.

Tone alert The ALU tells the pager to sound a tone when a page is received.

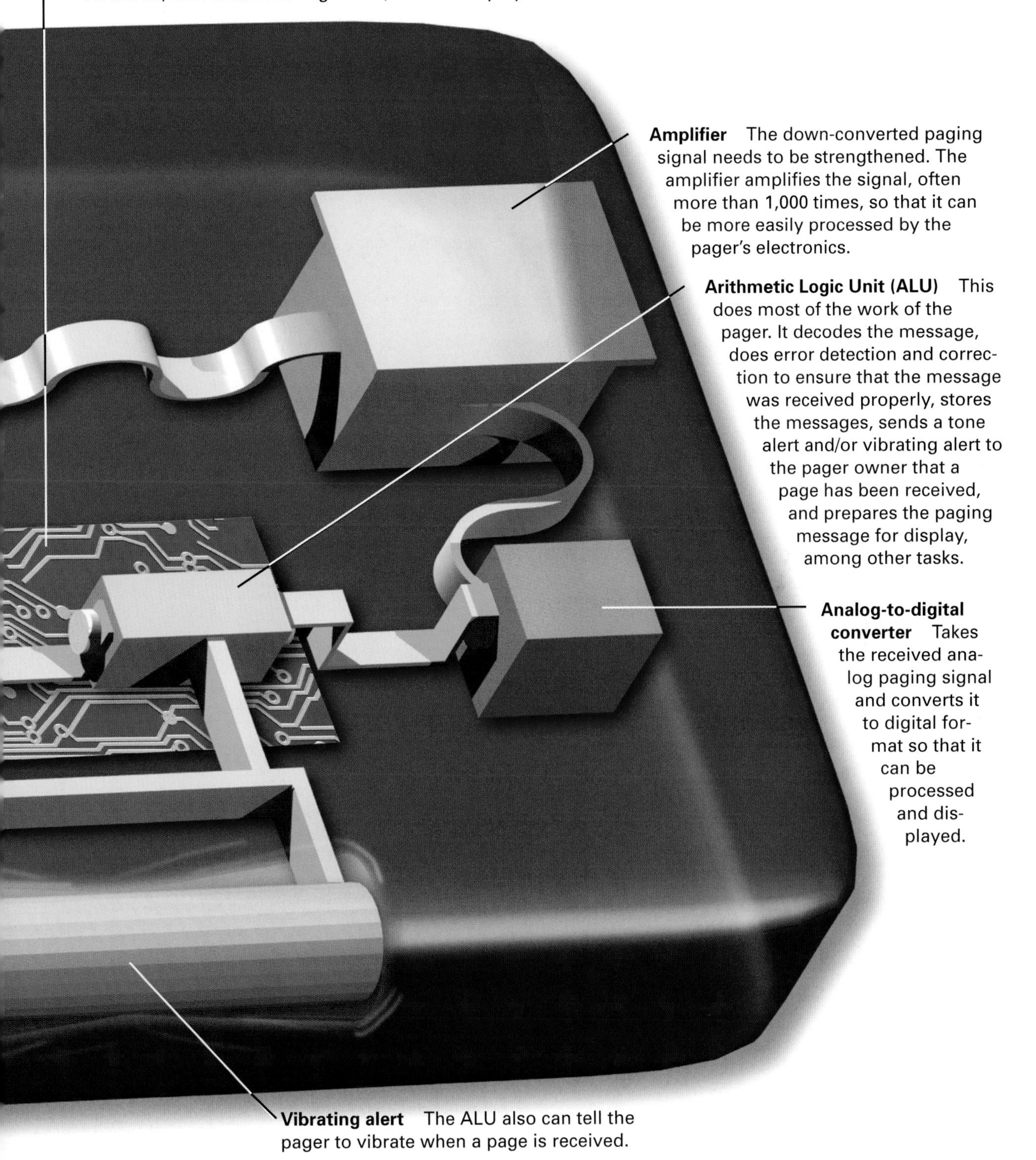
Microcontroller This contains much of the brains of the pager and does the processing, conversion, and display of the paging signal. There are three components to it: the analog-to-digital converter, the Arithmetic Logic Unit, and the display driver.
Amplifier The down-converted paging signal needs to be strengthened. The amplifier amplifies the signal, often more than 1,000 times, so that it can be more easily processed by the pager's electronics.
Arithmetic Logic Unit (ALU) This does most of the work of the pager. It decodes the message, does error detection and correction to ensure that the message was received properly, stores the messages, sends a tone alert and/or vibrating alert to the pager owner that a page has been received, and prepares the paging message for display, among other tasks.
Analog-to-digital converter Takes the received analog paging signal and converts it to digital format so that it can be processed and displayed.
Vibrating alert The ALU also can tell the pager to vibrate when a page is received.

How Pagers Work

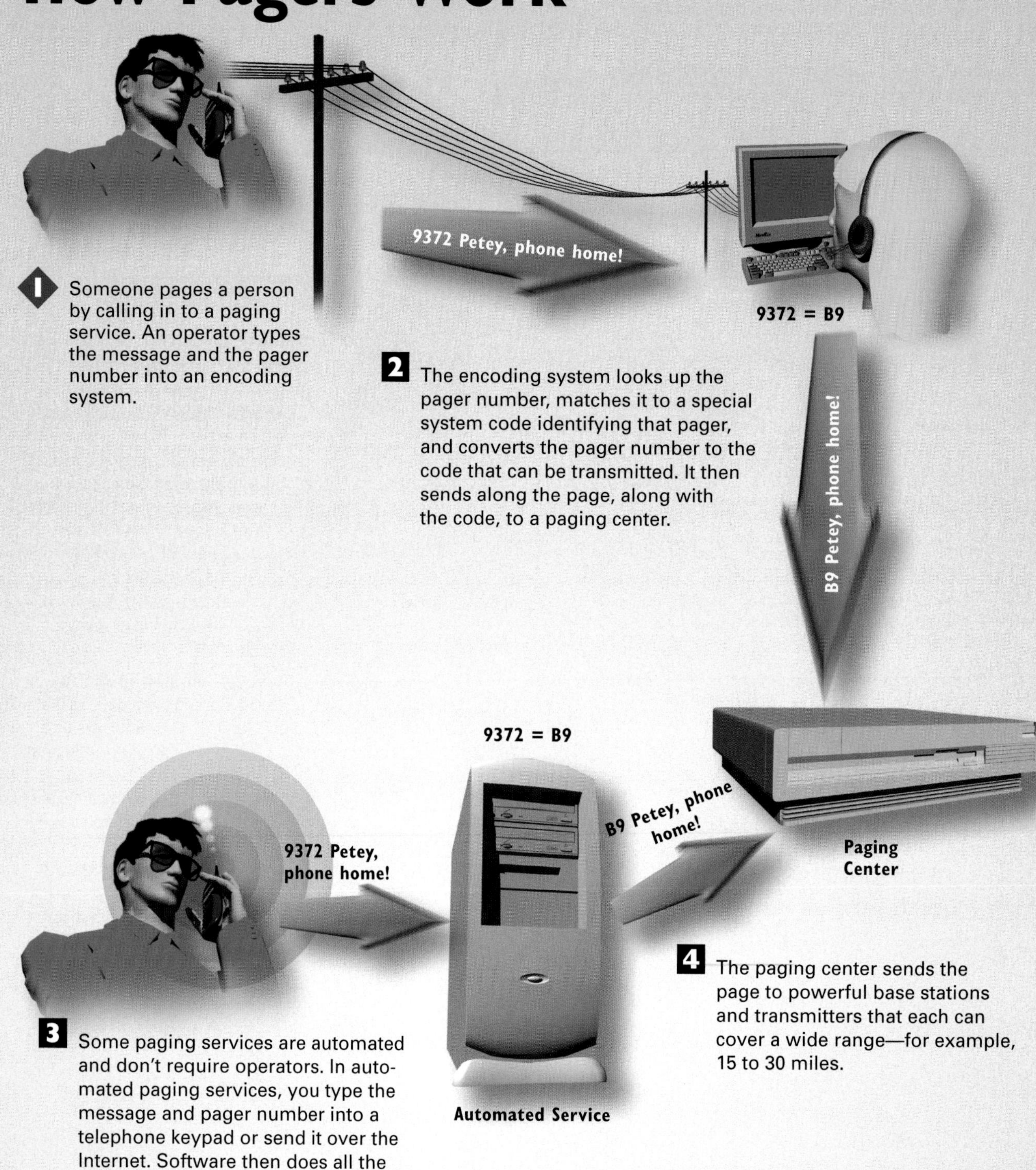
9372 Petey, phone home!
9372 = B9
B9 Petey, phone home!
9372 = B9
9372 Petey, phone home!
B9 Petey, phone home!
Paging Center
Automated Service

1. Someone pages a person by calling in to a paging service. An operator types the message and the pager number into an encoding system.

2. The encoding system looks up the pager number, matches it to a special system code identifying that pager, and converts the pager number to the code that can be transmitted. It then sends along the page, along with the code, to a paging center.

3. Some paging services are automated and don't require operators. In automated paging services, you type the message and pager number into a telephone keypad or send it over the Internet. Software then does all the work that a manual operator would normally do.

4. The paging center sends the page to powerful base stations and transmitters that each can cover a wide range—for example, 15 to 30 miles.

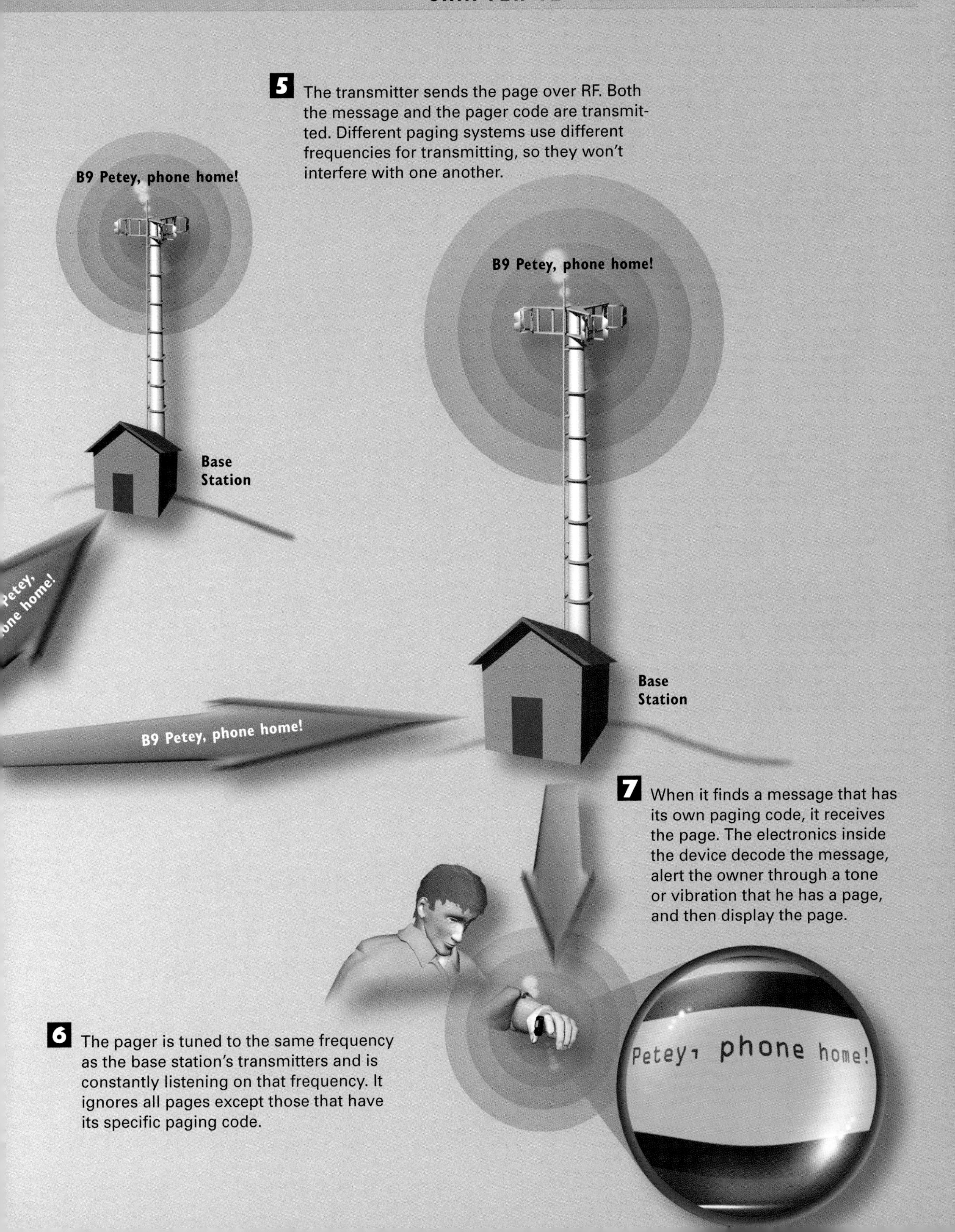
B9 Petey, phone home!
B9 Petey, phone home!
Base
Station
Base
Station
Petey, phone home!
B9 Petey, phone home!
Petey, phone home!

5 The transmitter sends the page over RF. Both the message and the pager code are transmitted. Different paging systems use different frequencies for transmitting, so they won't interfere with one another.

6 The pager is tuned to the same frequency as the base station's transmitters and is constantly listening on that frequency. It ignores all pages except those that have its specific paging code.

7 When it finds a message that has its own paging code, it receives the page. The electronics inside the device decode the message, alert the owner through a tone or vibration that he has a page, and then display the page.

How Two-Way Pagers Work

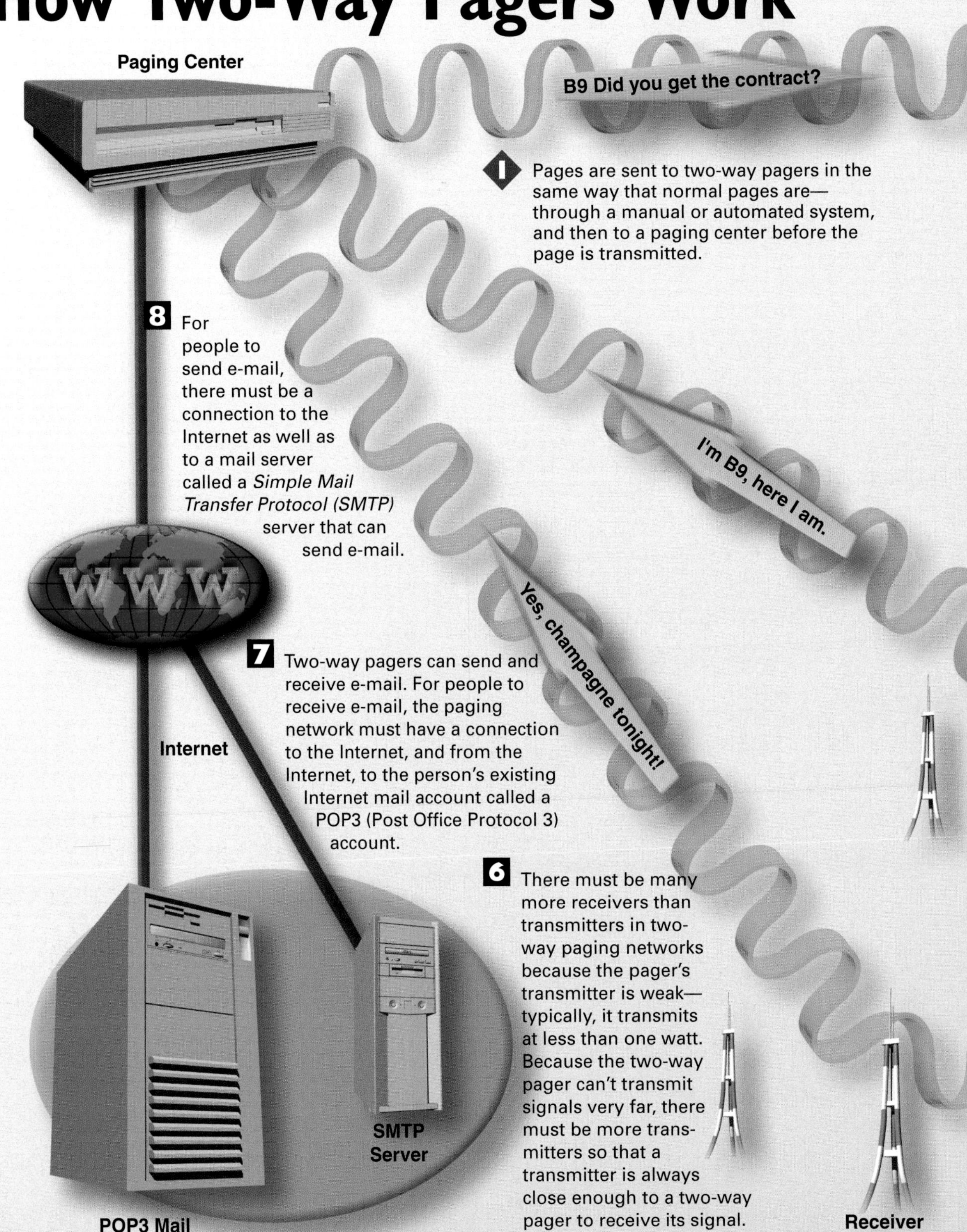
Paging Center
B9 Did you get the contract?
I'm B9, here I am.
Yes, champagne tonight!
Internet
SMTP Server
POP3 Mail Server
Receiver

1. Pages are sent to two-way pagers in the same way that normal pages are—through a manual or automated system, and then to a paging center before the page is transmitted.

6. There must be many more receivers than transmitters in two-way paging networks because the pager's transmitter is weak—typically, it transmits at less than one watt. Because the two-way pager can't transmit signals very far, there must be more transmitters so that a transmitter is always close enough to a two-way pager to receive its signal.

7. Two-way pagers can send and receive e-mail. For people to receive e-mail, the paging network must have a connection to the Internet, and from the Internet, to the person's existing Internet mail account called a POP3 (Post Office Protocol 3) account.

8. For people to send e-mail, there must be a connection to the Internet as well as to a mail server called a *Simple Mail Transfer Protocol (SMTP)* server that can send e-mail.

Base
Station
B9 Did you get the contract?

3 Using the location information, the network has only one base station and transmitter send out a page to the pager. There's no need to clog up the rest of the network with the page, because the network knows where the pager is located. This helps network operators cut down on unnecessary network traffic.

4 The two-way pager receives and displays the page in the same way that a one-way pager does.

Wireless Tidbit

Two-way paging systems are becoming increasingly popular for people who want to be able to receive e-mail the instant it's received, and be able to send e-mail as well. The most popular device for doing this is the Blackberry, which has become *de rigueur* for many in the high-tech set.

2 Two-way pagers register themselves with the paging network, so the network always knows where each individual two-way pager is located.

5 Two-way pagers include keypads so that messages can be answered or e-mail can be composed. Rather than having only a receiver, as in normal pagers, they have transceivers that can send and receive FR signals. So, messages and e-mails can be composed on the two-way pager and then transmitted.

Yes, champagne tonight!
Did you get the contract?
Yes!!! Champagne Tonight!

P A R T

Understanding Wireless Networks

A time is coming—it's coming soon, and is already here for many people—when wherever you go, you'll be connected to a network. A computer network, that is.

Most of us are used to the idea of being connected to a telephone or paging network wherever we go. With some exceptions and dead spots, when you carry your cell phone around, all you need to do to connect to the world, or see whether anyone wants to connect to you, is turn on the power. Worldwide communications are only a few button pushes away. And the same holds true for paging. Wherever you go, your little beeping companion will always let you know when someone needs to get in touch with you.

That's not the case, however, when it comes to computers. If you happen to carry a laptop, you can't get instant communications. You'll have to be sure you have a modem, then find a nearby telephone, and then dial numbers and hope for the best. For many people, "mobile computing" isn't very mobile—it's tethered by a phone line and access to a landline telephone.

Your computer is tethered even inside a corporation or your home. If you want to be connected in a corporate network, you must be near a jack that can get you into your company's network through an Ethernet cable. At home, you'll need to be near a telephone, or if you connect to the Internet through a cable modem or DSL line, you'll have to be near those cables or lines.

In short, computer communications has a long way to go when it comes to wireless network and Internet access.

That's all changing, however. Wireless networks have arrived, and although they certainly aren't everywhere yet, they're becoming more popular. One day, they'll become very common, and you'll be able to roam with a computer wherever you want, and be instantly connected to a network or the Internet.

In this section of the book, we'll take a look at wireless networks, both at home and in the office. We'll look at their inner workings and how they'll be used in a typical home or office. And we'll look at the workings of the two most popular kinds of wireless networks—Bluetooth and the family of IEEE 802.11 standards.

In Chapter 13, "How a Wireless Network Works," we'll look at the business use of wireless networks and see how a typical corporate wireless network works. We'll see how hubs and routers handle communications, examine how computers connect to the network, see how corporate offices connect to one another, see how traveling employees can connect to the home network wirelessly, and more. We'll see that corporate networks truly will lead to the completely mobile worker, both in the office and on the road. In this chapter, we'll also take a look at the inner workings of a wireless network card and see how it's able to let computers make connections to wireless networks.

Chapter 14, "How Home Wireless Networks Work," looks at wireless networking at home. Surprisingly, wireless networking has come to the home before it has come to the corporate world. You can buy wireless networks inexpensively—for several hundred dollars—that will connect all the computers in a home. Typically, these networks primarily are used to allow several computers to share a high-speed Internet connection, such as a cable modem or DSL modem. But they also allow printers and other devices to be shared, as well as allowing computers to share files with one another. These networks are quite easy to set up and are especially helpful for people who don't want to have to string Ethernet cabling throughout a home. In this chapter, we'll also look at a home network that goes beyond computers—one that can include radios and appliances such as microwave ovens in a wireless home network attached to the Internet.

In Chapter 15, "Understanding Bluetooth and IEEE 802.11 Networking," we'll look closely at the two most popular wireless network standards. Bluetooth is named after Harald Blatand, King of Denmark from approximately 940 to 985, who united Denmark and Norway. Bluetooth technology is designed to allow many different kinds of devices to talk to one another, from computers to cell phones to stereos—in fact, just about any device you can name. It's designed as a peer-to-peer network that allows these devices to talk directly to one another, without extra hardware such as network servers. The IEEE 802.11 networking standard, on the other hand, is designed specifically for computers, and although it also can work as a peer-to-peer network, it most often will be used in concert with traditional networking hardware such as hubs, routers, and servers.

CHAPTER

13

How a Wireless Network Works

THE lifeblood of businesses is information; its heart is communication. The more information it has, and the better it shares that information, the healthier it is.

The primary way that companies get and share information today is through their *local area networks (LANs)*. LANs serve two purposes: They allow people in the business to share information with each other and to get at corporate information and resources; and they allow people to access the Internet, through the LAN's connection to the Internet through hardware called routers.

Just about any corporation you walk into has LANs with Internet access. By now, that's old hat, and pretty much a requirement for any company to do business. What you're just starting to see are wireless LANs that allow people to connect to the corporate network and the Internet without wires. There are many benefits for this for companies, not the least of which is that it makes it far easier for people to get at information and share it, than when they use a traditional LAN. For example, in a company with wireless LAN, wherever people go with their laptops, they're automatically connected to the LAN. So, people can take laptops into meetings with them, into other people's offices, and get all the information and corporate resources they want.

It also means that people who have wireless *personal digital assistants (PDAs)*, like the Palm, also can get immediate access to the network wherever they go.

Wireless networks also let corporations extend their LANs far beyond the walls of a single building. Wireless point-to-point networks, called *wireless bridges*, let companies who have buildings with clear lines of sight to each other send network data over microwave transmitters and receivers. Some companies that span the U.S. and Mexican border use these wireless bridges as a way to have corporate offices communicate with each other, and so get around the steep cost of international phone calls.

Wireless networks can extend the corporation's reach to its employees in other ways, as well. People can connect to the network with wireless devices using a *Virtual Private Network (VPN)* over a wireless gateway, and get at their e-mail and other corporate resources even when they're outside the office. Because the VPN uses encryption, no one can steal data being sent and received.

Wireless networks are still somewhat rare in the business world, and even when they become more popular, they'll rarely be wireless-only. Instead, wireless networks will be tacked on to existing wired networks through the use of wireless access points. These wireless access points allow computers with wireless network cards to hop onto the LAN. In the next two illustrations, you'll see how corporate networks work, and you'll also take a look inside a wireless network card.

How a Corporate Wireless Network Works

Server

Router

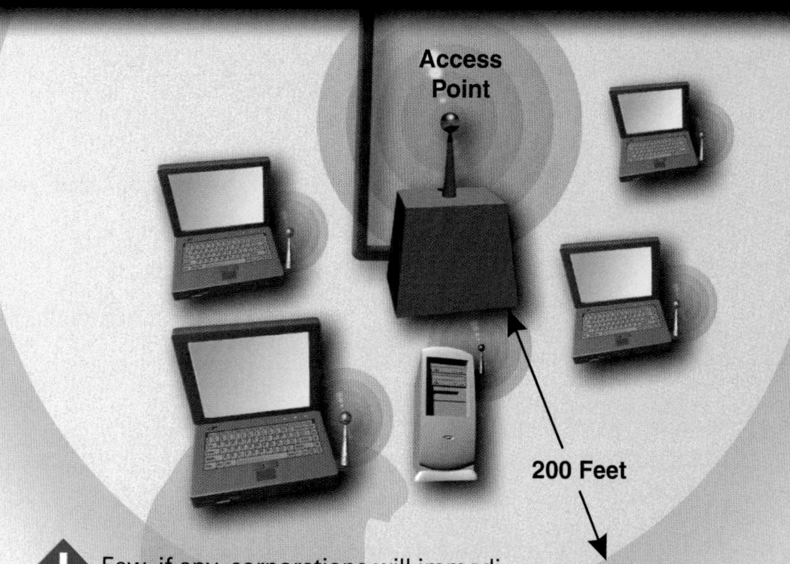

1 Few, if any, corporations will immediately go to all-wireless networks. More likely, they'll combine a wireless network with their existing Ethernet-based local area network (LAN). (And they most likely will choose some form of an 802.11 wireless network. For more information about 802.11 networks, turn to Chapter 15, "Understanding Bluetooth and IEEE 802.11 Networking.") In this kind of configuration, the network servers and the routers for connecting the network to the Internet, will all be on the existing wired network.

2 Wireless access points connect to the existing LAN, giving nearby computers with wireless network cards access to the network and, through it, the Internet.

3 The exact amount of space that each access point covers varies according to the layout of the building, type of wireless connection, and any obstacles. Typically, however, access points can allow computers within a 100- or 200-foot radius to send and receive information to the network wirelessly.

4 Access points can be spread out over a floor, a building, or many buildings, extending the network to many parts of the corporation.

Mail
Server

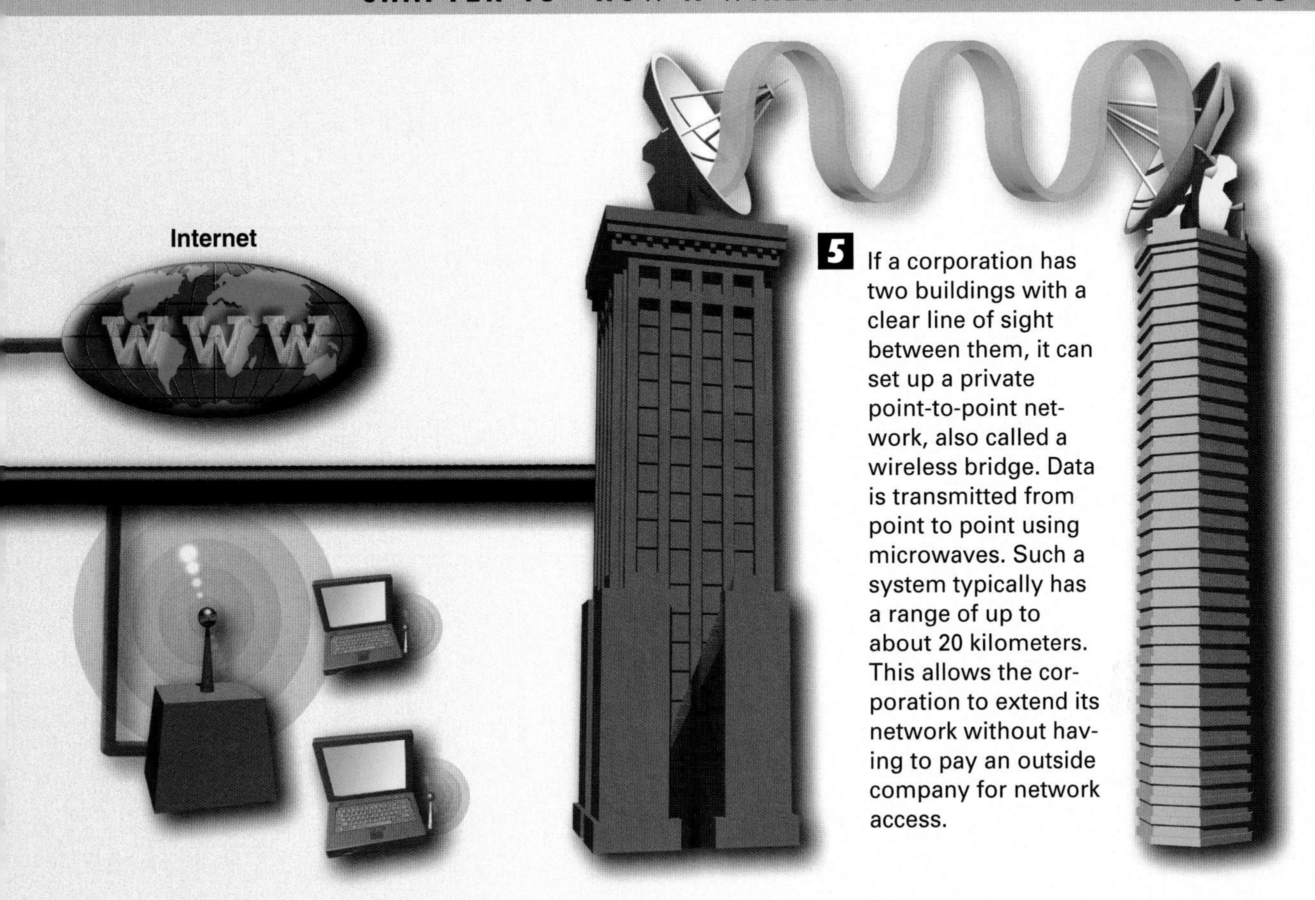

5 If a corporation has two buildings with a clear line of sight between them, it can set up a private point-to-point network, also called a wireless bridge. Data is transmitted from point to point using microwaves. Such a system typically has a range of up to about 20 kilometers. This allows the corporation to extend its network without having to pay an outside company for network access.

6 Some corporations allow their workers to access corporate information, such as e-mail boxes, remotely with cellular devices, such as personal digital assistants and wirelessly equipped laptops. There are several ways this can be done, but one way allows the workers to connect to a server using a wireless Virtual Private Network (VPN), which automatically encrypts all data so that no one can snoop on it.

How a Wireless Network Card Works

1. The Boot ROM (Read-Only Memory) and system memory area handles startup routines when the computer is turned on with the card inside it. It also contains the basic instructions for operating the card. In this example, it contains instructions for a wireless PCMCIA card, which is a credit-card–size wireless device that fits into a special PCMCIA slot in a laptop computer. Wireless network cards can connect to computers in several different ways. In the case of desktop computers, the wireless card often is put into a free slot inside the computer, right on the bus. For laptop computers, it usually is connected through a PCMCIA slot. Both laptops and desktops also can use USB wireless cards, which plug into the USB port. Pictured here are a PCMCIA card and the PCMCIA connector.

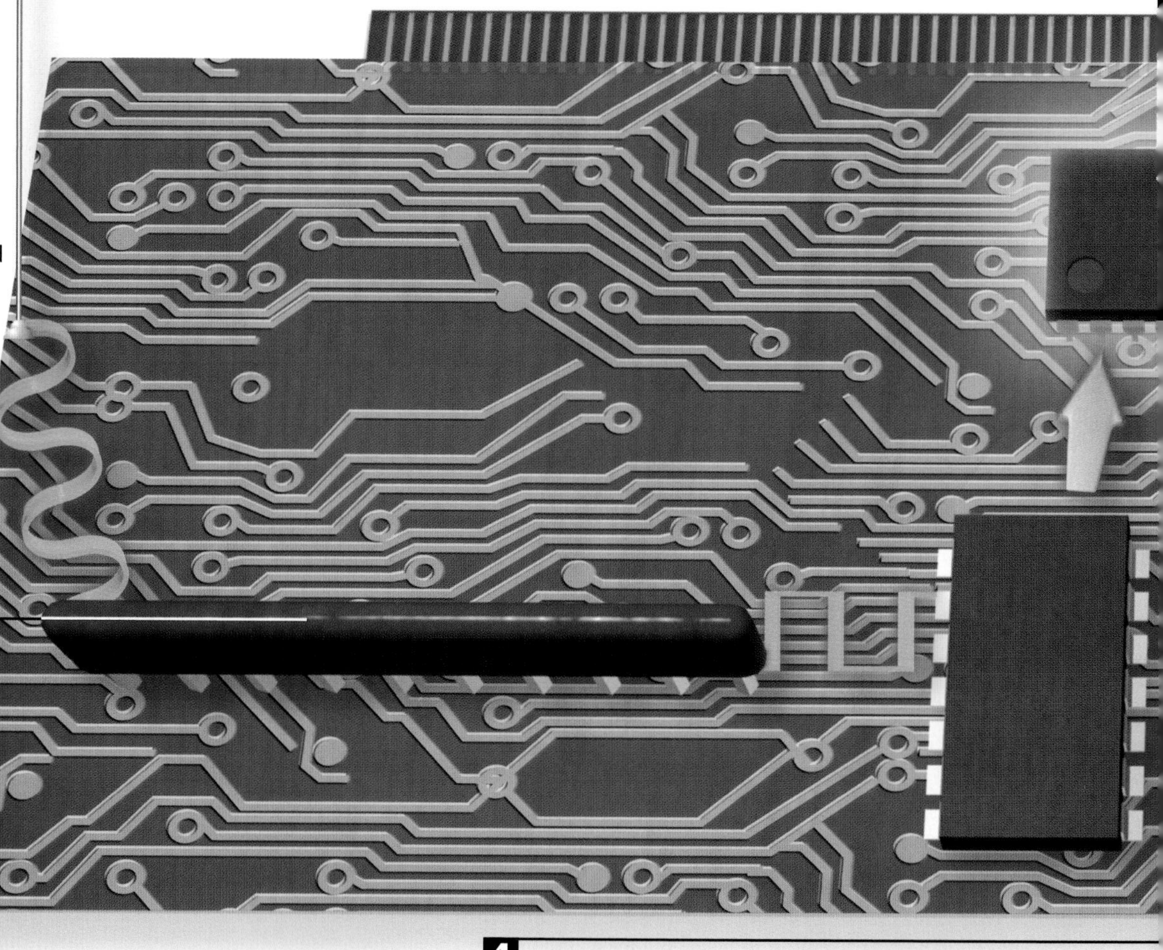

Antenna

2. The card has a small antenna through which it sends and receives information to and from the wireless base station.

3. A radio transceiver is connected to the antenna. The transceiver handles the job of modulating information from the computer onto RF waves, and of demodulating information received from the antenna into digital signals that the PCMCIA card and the computer can understand. It can both receive and send information wirelessly.

4. The brains of the card is the controller. It takes the data from the transceiver, processes it, and does the work of being the interface between the network and the computer. It also shuttles data into memory when necessary. After receiving information from the radio, it processes it so that the computer can understand it, and sends it to the card's PCMCIA interface.

5 The PCMCIA interface sends the data into the computer. When information is instead sent from the computer to the network, the entire process is reversed—the data starts in the computer, gets sent through the PCMCIA interface to the card's controller, then to the transceiver, and then out from the antenna to the network.

CHAPTER 14 How Home Wireless Networks Work

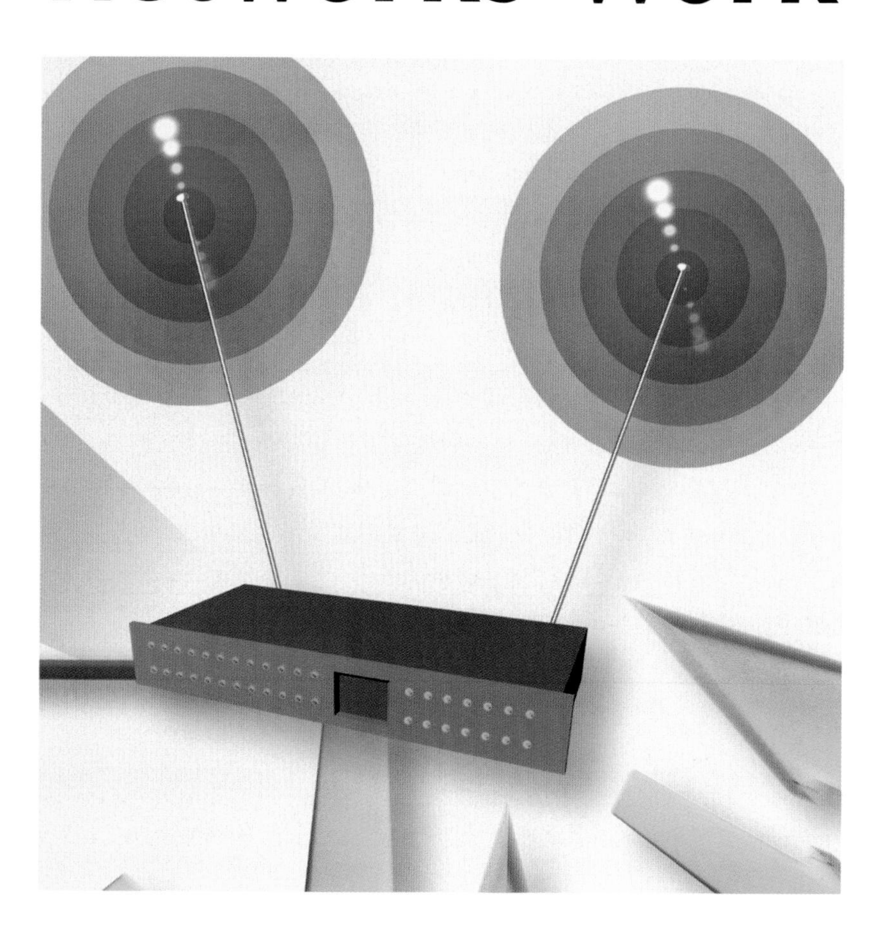

WIRELESS networks aren't only for business. In fact, as of now, they're probably used more in the home than in businesses. There are several reasons for that. A big one has to do with cost. Corporations are located in office buildings that are already wired—they have strung through them the Ethernet cables that connect computers to the network, so building an entirely new wireless network is a very expensive proposition.

By way of contrast, homes don't have Ethernet cables strung through the walls. Therefore, to network computers in several rooms—in a study, a home office, and several children's bedrooms, for example—one would have to snake cables through the walls, and that costs a significant amount of time and money. With a wireless network, you don't need to spend that time and money.

The other reason that wireless networks have become popular at home is that the simple ones used in homes are very easy to set up, and have become relatively inexpensive, costing only several hundred dollars. To build a wireless network at home, you can buy a network kit with all the required pieces. You'll need a wireless hub/router that connects all the computers to one another and to the Internet. And you'll need to buy wireless network cards for each computer you want to connect to the network. The computers all connect to the wireless hub/router, and the hub/router routes all the traffic between the computers and between the computers and the Internet.

For computers to be networked this way, they all must have what are called *IP addresses*. When a computer has an IP address, it can get full access to the Internet and to other computers on the network. Among other jobs, the hub/router assigns IP addresses to all computers in the network.

The main reason why people install wireless networks is to share a high-speed Internet connection, such as a cable modem or a DSL modem. But they can use the network for other things as well, notably sharing devices such as printers, to send files back and forth between computers, or to play computer games over the network against other family members.

Although today, mainly computers at home are networked wirelessly, in the future other kinds of devices and appliances will be connected to one another as well, such as radio receivers; small, inexpensive e-mail devices that only send and receive e-mail; and even traditional home appliances such as refrigerators, microwave ovens, and alarm clocks. Not only will they connect to one another, they'll connect to the Internet as well.

Connecting these kinds of devices and appliances will make life more convenient—you'll be able to use your refrigerator to automatically generate shopping lists, for example, and send orders directly to grocery stores. And you'll have an alarm clock that can change the time it awakens you based on traffic reports it garners from the Internet. These kinds of devices aren't mere fantasy—they are already being sold or tested, and soon will be sold in a department store near you. Initially, many will require wires to connect to each other and the Internet, but soon they'll connect wirelessly as well, and some already do. (You can find more information about Internet appliances at `www.thaliaproducts.com`.)

How a Home Wireless Network Works

1 The main reason why people set up wireless home networks is to share high-speed Internet access, such as through a cable modem or DSL modem, among several computers. They choose a wireless network rather than a wired one in large part because of the expense and difficulty in running wires through the walls of their homes. If a home wireless network will be used to give computers access to the Internet, a device called a *wireless hub/router* must be attached through an Ethernet cable to the cable modem or DSL modem.

3 The wireless hub/router does two primary jobs: It connects all the PCs to each other so that they can share files and devices such as printers, and it connects all the PCs to the Internet, so that they each can have a high-speed Internet connection. For the hub/router to do its job, it needs what's called an *IP address*. An IP address allows a device to get on to the Internet. The IP address is given to the hub/router by a server run by the Internet provider that runs the cable or DSL service.

7 In addition to gaining Internet access through the hub/router, the PCs also can share resources such as printers. So, any computer on the network can print to a printer attached to any other computer, as long as the computes are set up to share resources.

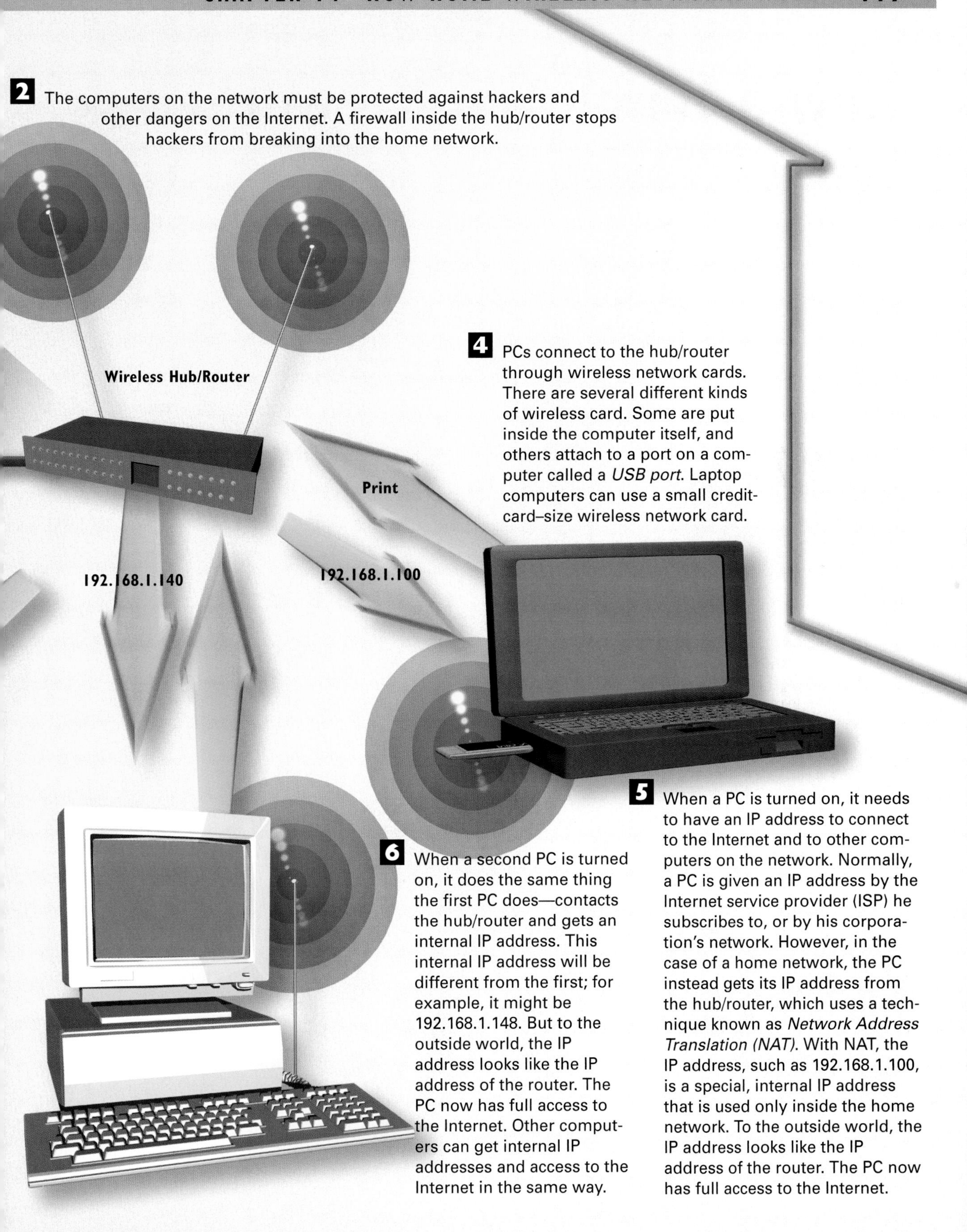

2 The computers on the network must be protected against hackers and other dangers on the Internet. A firewall inside the hub/router stops hackers from breaking into the home network.

4 PCs connect to the hub/router through wireless network cards. There are several different kinds of wireless card. Some are put inside the computer itself, and others attach to a port on a computer called a *USB port*. Laptop computers can use a small credit-card–size wireless network card.

5 When a PC is turned on, it needs to have an IP address to connect to the Internet and to other computers on the network. Normally, a PC is given an IP address by the Internet service provider (ISP) he subscribes to, or by his corporation's network. However, in the case of a home network, the PC instead gets its IP address from the hub/router, which uses a technique known as *Network Address Translation (NAT)*. With NAT, the IP address, such as 192.168.1.100, is a special, internal IP address that is used only inside the home network. To the outside world, the IP address looks like the IP address of the router. The PC now has full access to the Internet.

6 When a second PC is turned on, it does the same thing the first PC does—contacts the hub/router and gets an internal IP address. This internal IP address will be different from the first; for example, it might be 192.168.1.148. But to the outside world, the IP address looks like the IP address of the router. The PC now has full access to the Internet. Other computers can get internal IP addresses and access to the Internet in the same way.

How a Wirelessly Networked Home Works

Security and monitoring system Security systems can be connected to the home network and the Internet so that you can, for example, look through security cameras when you're far away and be sure that your house is okay. You also can put a Webcam in your young children's rooms and be able to monitor them when you're in another room or away from home.

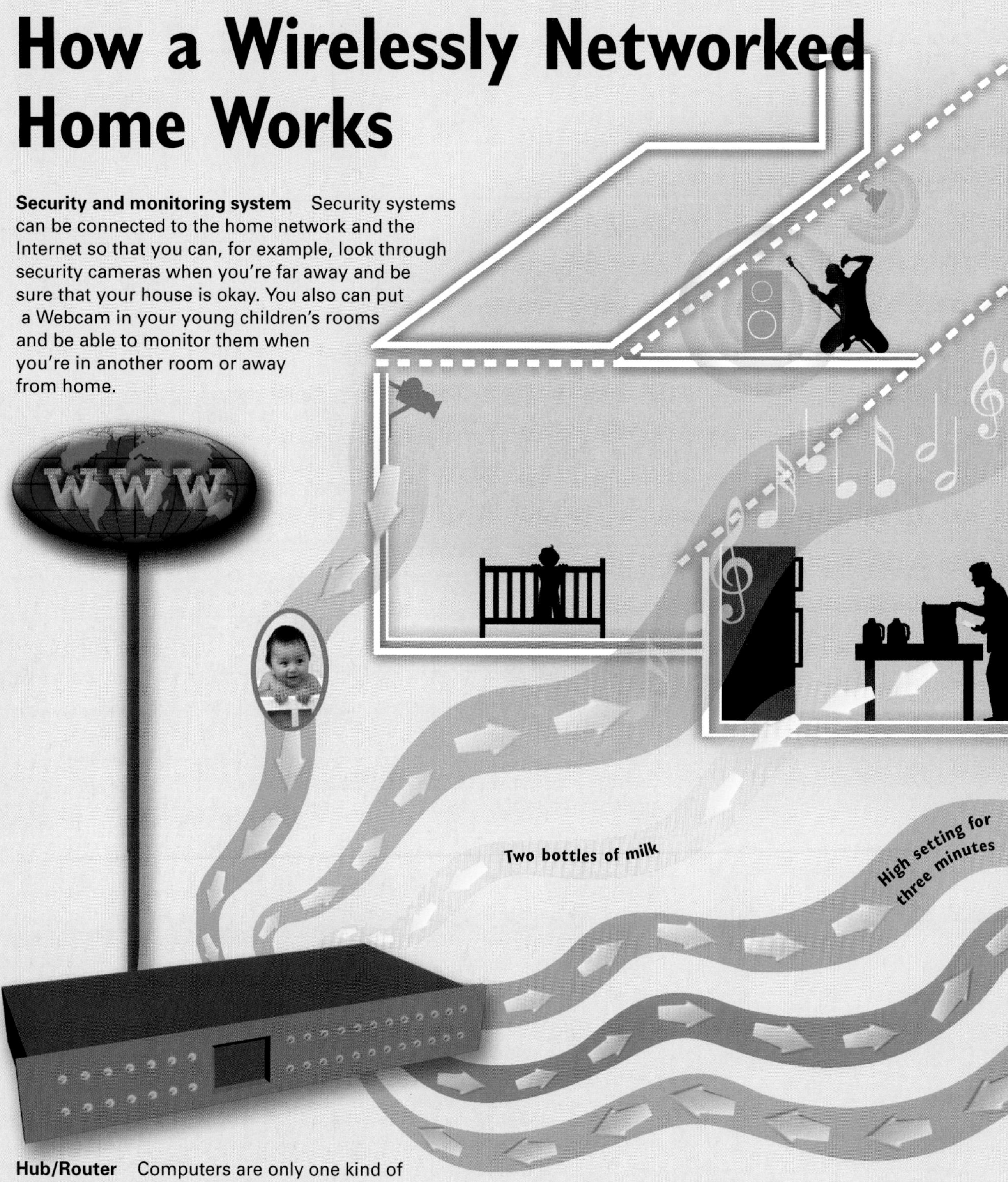
WWW
Two bottles of milk
High setting for three minutes

Hub/Router Computers are only one kind of home device that can be networked wirelessly. In theory, any kind of device or appliance—from your television to your heating system to your refrigerator—can be wirelessly networked and connect to the Internet. But no matter what devices are in the network, a wireless hub/router generally will connect them all to one another, and to the Internet.

Internet home control Ultimately, every part of your home will be able to be monitored and controlled from a home network, from the lights to the heating system, security system, and beyond. You'll be able to monitor and control your home either through a central keypad and monitor at home, through a PC at home, or through a PC or some other Internet-connected device when you're away from home. Sunbeam, which makes the Mr. Coffee coffeemaker, has announced that it will be selling a set of nine Internet-connected home devices, including a smart, Internet-connected alarm clock called the TimeHelper. The device can beep a warning when the Internet-connected Smart Coffeemaker runs low on water, can check Internet traffic reports to see what traffic is like and set your alarm to the best time for getting into work, and can even turn off or on your electric blanket based on weather reports. The Internet alarm clock is expected to sell for between $40 and $80.

Internet radio tuner One of the newest home devices that uses an Internet connection is a radio tuner. These tuners connect to the Internet and play Internet radio stations. They can play the music in speakers throughout the house. Those speakers can be wired directly to the tuner in the traditional way, or can be connected wirelessly, or through a home's telephone wires.

E-mail terminal For people who want access to e-mail throughout the house but can't afford a PC in every room, inexpensive e-mail terminals can be used. These terminals are used primarily for sending and receiving e-mail, although some have basic Web-browsing capabilities as well.

Heavy traffic
Hi, Mia! How are you?

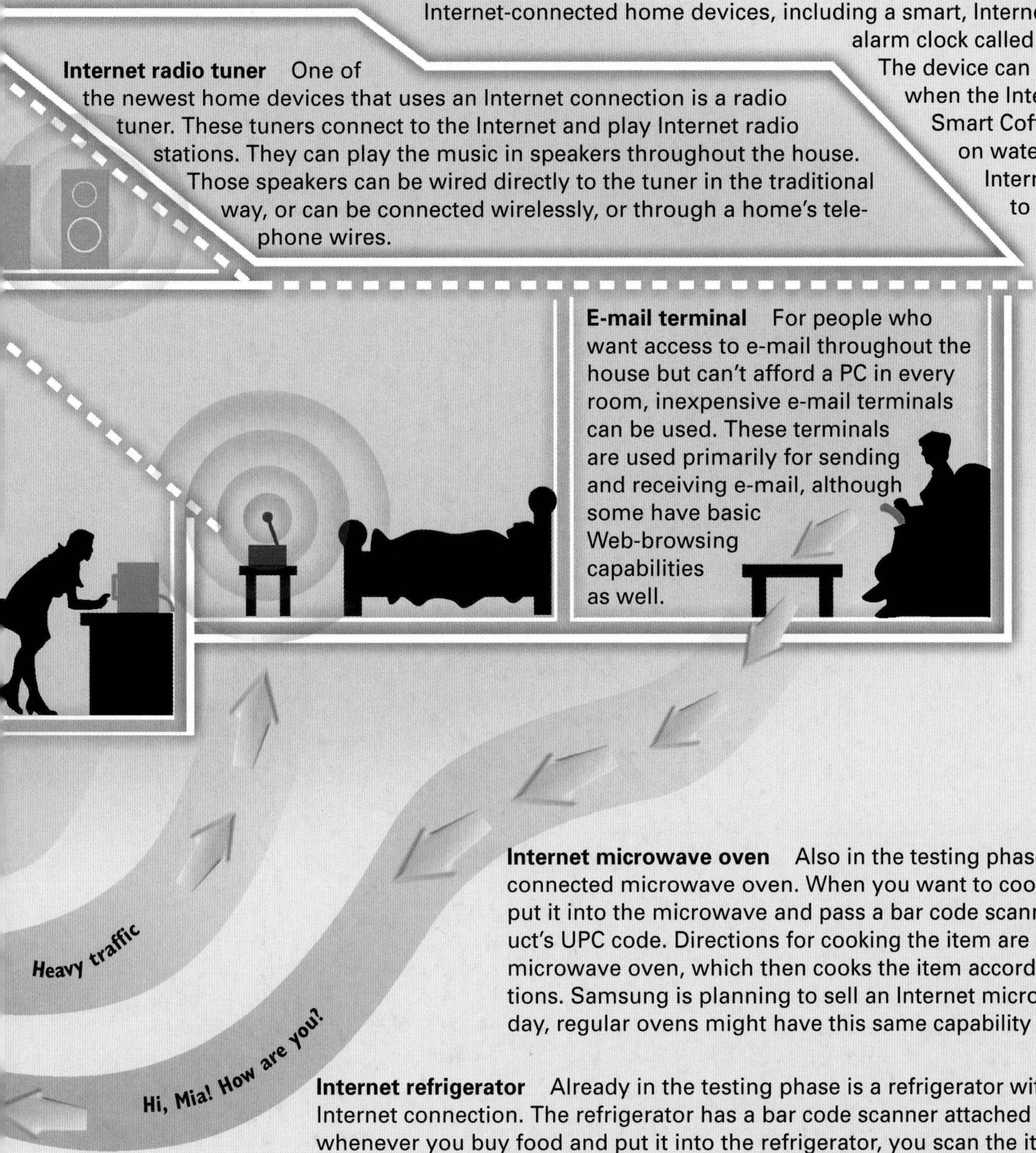

Internet microwave oven Also in the testing phase is an Internet-connected microwave oven. When you want to cook something, you put it into the microwave and pass a bar code scanner over the product's UPC code. Directions for cooking the item are sent directly to the microwave oven, which then cooks the item according to the directions. Samsung is planning to sell an Internet microwave oven. Some day, regular ovens might have this same capability as well.

Internet refrigerator Already in the testing phase is a refrigerator with an Internet connection. The refrigerator has a bar code scanner attached to it, and whenever you buy food and put it into the refrigerator, you scan the item's UPC code. The refrigerator keeps a record of what you've bought and can create automatic shopping lists based on your purchases. You then can send the shopping list from the refrigerator directly to a grocery store, which could then deliver the goods to your home. Refrigerator makers Electrolux, GE, and Whirlpool are planning to sell these appliances. Initially, refrigerators and other kitchen and home devices might need wires to connect to the Internet, but soon wireless connections will follow.

CHAPTER

15

Understanding Bluetooth and IEEE 802.11 Networking

WIRELESS networks are new enough that there's no single, dominant established standard yet for how computers and other devices connect to each other and to the Internet. It probably will be years before a dominant standard is established—and it might well be that no single standard will emerge, and that several will be in widespread use.

For now, there are two main standards for wireless networking: one called Bluetooth, and the other known by the unromantic name IEEE 802.11. (It's pronounced "eye-triple-e-eight-o-two-dot-eleven. Often, though, the IEEE isn't used in referring to it.) Although they compete with each other, they actually are used for somewhat different purposes.

Bluetooth is designed to be used in devices of all sorts—everything from your computer to your stereo system and anything in between. The capability to network wirelessly with Bluetooth is built into a chip, and the chip is put into a device. Because of that, it is expected to become a low-cost way to network various devices. The technology was devised so that you need not do anything to hook a device into a network. Simply turn the device on, and it automatically looks around for another Bluetooth device. If it finds one or more, they set up wireless communications by themselves.

Bluetooth is an ad hoc network, which means that not only do the devices find each other on their own, but they can communicate directly with each other without having to go through a central device, such as a server or a network access point. This kind of network, when devices connect directly to one another, is known as a *peer-to-peer network.*

Bluetooth allows computers, telephones, *personal digital assistants (PDAs)*, and even home devices such as stereos and TVs to communicate with one another. Bluetooth devices automatically find each other, without your having to install them, or even ask them to find other devices. You also can use Bluetooth devices to access the Internet, as long as one of the devices is directly connected in some way to it.

Bluetooth networks can't be very large; if there are too many devices on one that try to communicate with one another, the network and devices can crash. Because of that, and because Bluetooth wasn't devised only for computers, another kind of wireless network has become popular—802.11. This kind of network is well-suited for working with the Ethernet local area networks popular in many corporations.

Although 802.11 can work in an ad hoc peer-to-peer manner, it primarily is used more like a traditional corporate network. Computers equipped with 802.11 network cards communicate with wireless access points, which connect the computers to the network. Often, 802.11 wireless networks connect to larger corporate networks. So, it's an ideal way for someone with a laptop computer to hook into a company network—no matter where the person goes with a laptop, immediate network access is available. Because it's often used in a corporate environment, it can use security services such as for encryption and authentication using the *Wired Equivalent Privacy (WEP)* protocol.

Both 802.11 and Bluetooth networks can be used either at home or in corporations. And, in fact, there's no reason why Bluetooth and 802.11 networks can't exist side by side. Bluetooth is mainly for many kinds of devices, including those used for home entertainment, whereas 802.11 is best suited for computers. So, a Bluetooth network could be used for entertainment and similar reasons (even though it's also suitable for data transmissions), whereas the 802.11 could be used for computers and Internet access.

How Bluetooth Works

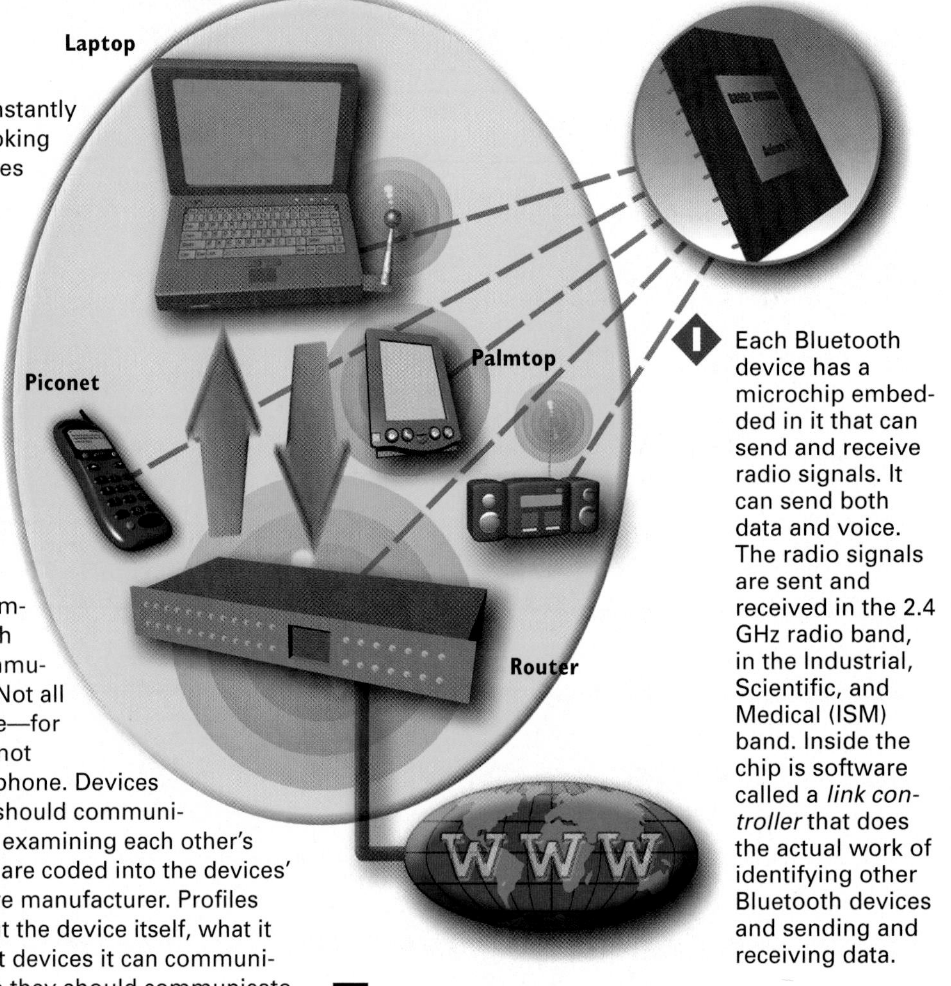

Laptop

Piconet

Palmtop

Router

1 Each Bluetooth device has a microchip embedded in it that can send and receive radio signals. It can send both data and voice. The radio signals are sent and received in the 2.4 GHz radio band, in the Industrial, Scientific, and Medical (ISM) band. Inside the chip is software called a *link controller* that does the actual work of identifying other Bluetooth devices and sending and receiving data.

2 The Bluetooth device constantly sends out a message, looking for other Bluetooth devices within its range.

3 When a Bluetooth device finds another device, or more than one device, within its range, they go through a series of communications that establish whether they should communicate with one another. Not all devices will communicate—for example, a stereo might not communicate with a telephone. Devices determine whether they should communicate with one another by examining each other's Bluetooth "profiles" that are coded into the devices' hardware by the hardware manufacturer. Profiles contain information about the device itself, what it is used for, and with what devices it can communicate. If devices determine they should communicate with one another, they establish a connection. The connection of two or more Bluetooth devices is called a *piconet*.

4 When the connection is established, the devices can communicate with one another. You could use a Bluetooth device to access information from the Internet, if the device from which it's accessing the data is connected to the Internet. For example, you could have a home network with Bluetooth capability and connect a palmtop computer to the Internet by connecting through the home network.

5 If there are many Bluetooth devices or piconets near each other, their radio signals could conceivably interfere with one another. To be sure that doesn't happen, Bluetooth uses spread-spectrum frequency hopping. In this technique, the transmitters change their frequency constantly—1,600 times per second. In this way, the chance of interference is very small—and if there is interference, it will happen only for a tiny fraction of a second. When two or more devices are connected in a piconet, one device is the master and determines the frequencies to switch among. It instructs all the other devices on which frequencies to switch to, and when.

Wireless Tidbit

Bluetooth is named after Harald Blatand, king of Denmark from approximately 940 to 985. His last name, Blatand, roughly translates as "blue tooth." The king united Denmark and Norway, so it makes sense that a networking technology would be named after him—after all, networks are designed to unite people as well.

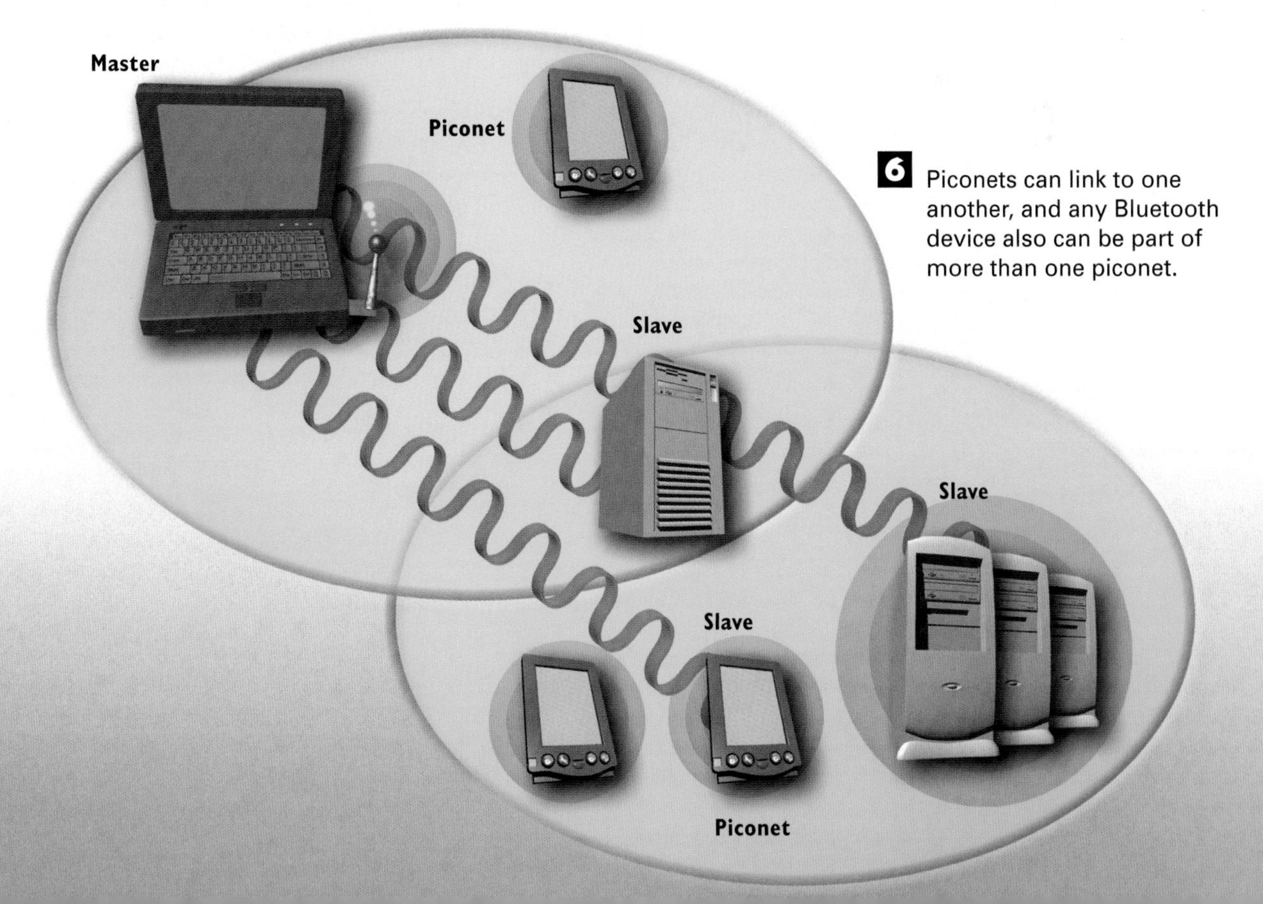

6 Piconets can link to one another, and any Bluetooth device also can be part of more than one piconet.

How IEEE 802.11 Networking Works

Ethernet

1 A key component of an 802.11 network is an *access point* (often called an AP). The access point consists of a radio transmitter and receiver as well as an interface to a wired network such as an Ethernet network, or directly to the Ethernet. The access point serves as a base station and a bridge between the wireless network and a larger Ethernet network or the Internet.

2 For a computer to become part of the network, it must be equipped with an 802.11-compatible wireless network card, so that it can communicate with the access point. Each computer that's part of the network usually is referred to as a *station*. Many stations can communicate with a single access point. An access point, along with all the stations communicating with it, are collectively referred to as a Basic Service Set (BSS).

3 When a station is first turned on or enters an area near the access point, it scans the area to look for an access point by sending out packets of information called *probe request frames* and waiting to see whether there is an answering probe response from a nearby access point. If the station finds more than one access point, it chooses one based on signal strength and error rates.

Probe Request Frame

4 Stations communicate with the access point using a method called *Carrier Sense Multiple Access with Collision Avoidance (CSMA/CA)*. It checks to see whether other stations are communicating with the access point, and if they are, it waits a specified random amount of time before transmitting information. Waiting a random amount of time ensures that the reattempts at transmission don't continuously collide with one another.

5 Before a station transmits information or a request, it first sends a short packet of information called a *Request to Send (RTS)*, which includes information about the request or data to come, such as its source, destination, and how long the transmission will take.

Basic Service Set

Can I talk? (RTS)

Go ahead (CTS)

Here's the data!

Got it! (ACK)

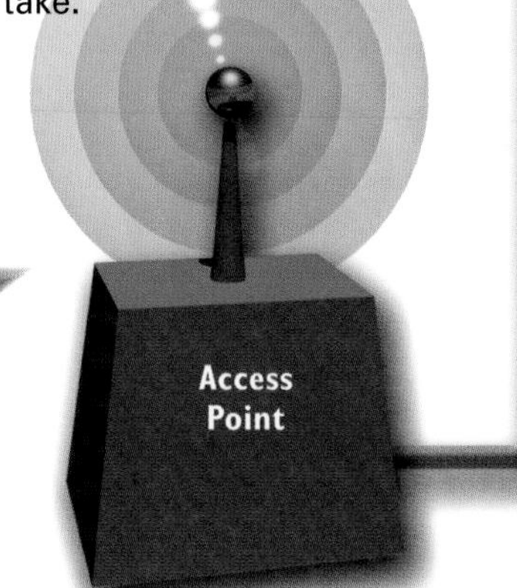

6 If the access point is free, it responds with a short packet of information called a *Clear to Send (CTS)*, telling the station that the access point is ready to receive information or requests.

7 The station sends the packet to the access point. After the packet is received, the access point sends an ACK (Acknowledgment) packet confirming that the data was received. If an ACK packet isn't sent, the station resends the data until it receives an ACK packet.

Ethernet

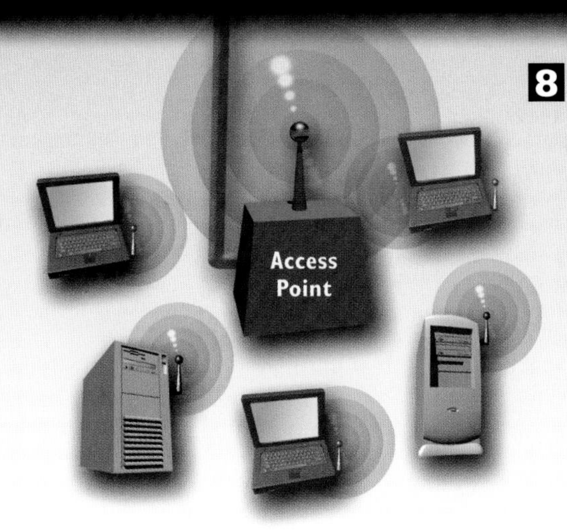

8 An 802.11 network can have many access points and many stations. Stations can move from access point to access point. Taken together, all the access points and stations are called an *Extended Service Set (ESS)*.

Access
Point

Wireless Tidbit

There are three standards for networking using the 802.11 wireless standard—the original standard, which allows for speeds of up to 2 Mbps; the 802.11b standard, which allows for transfer speeds of up to 11 Mbps; and the 802.11a, standard, which allows for transfer speeds of up to 54 Mbps. The 802.11 and 802.11b standards communicate in the 2.4 GHz spectrum, whereas the 802.11a communicates in the 5 GHz spectrum—a problem in Europe and Japan, because of potential interference with other devices that use that spectrum. At the moment, there are no serious potential problems for interference in the United States.

9 The 802.11 standard also allows stations to communicate directly with one another, without a connection to an access point, a network, or the Internet. When stations communicate directly with one another, it's called a *peer-to-peer network*. This allows the stations to do things such as share files and communicate directly with one another.

Extended Service Set

PART 5

The Wireless Internet

THE greatest revolution in wireless technology since the invention of the cell phone is, undoubtedly, the wireless Internet. In the same way that the cell phone gave people instant access to the global telephone network, the wireless Internet gives people instant access to the other great globe-scanning communications medium.

Most of us are used to accessing the Internet through a computer, but with the advent of the wireless Internet, there will be many different ways to gain access to everything from e-mail to Web pages and more. You can browse the Web or get e-mail using a cell phone, for example, or from a personal digital assistant (PDA) such as a Palm device. And there are many other ways that the Internet can be accessed wirelessly.

In this section of the book, we'll look at the wireless Internet. You'll learn everything from the basic underlying technologies that make the Internet possible, to the different ways that cell phones access the Internet, to how wireless PDAs work, and much more.

In Chapter 16, "Understanding the Internet," you'll learn about the basic technologies and protocols that make the Internet work. You'll start off with the basics, seeing how data moves through the giant data network. Then, you'll learn about the most basic of Internet protocols—TCP/IP, the Transmission Control Protocol/Internet Protocol. Together, these protocols do the job of delivering data across the world. You'll also learn how the Web works—how Web servers and your computer work together to let you visit any Web page in the world. The chapter also covers that ubiquitous type of communications—e-mail. You'll see how the Internet takes an e-mail you write on your computer (or cell phone, these days) and delivers it to mailboxes anywhere in the world.

Chapter 17, "How Cell Phones Access the Internet," delves into the mysteries of how your cell phone can do things such as browse the Web, grab Internet information, and send and receive e-mail. Cell phones get onto the Internet using a protocol called the Wireless Access Protocol (WAP), which works together with TCP/IP. You'll see how WAP enables you do to things such as browse the Web. An important part of WAP is the Wireless Markup Language (WML). This language is related to the HTML language that builds Web pages. WML allows people to build Web sites specifically suited for cell phones. A related technology is WMLScript, a scripting language that adds interactivity to WAP pages. So, you'll learn how that technology works as well. Finally, you'll see how cell phones can send and receive e-mail.

In Chapter 18, "How XML and Voice XML Deliver Internet Data," you'll learn about two related technologies: the Extended Markup Language (XML) and the Voice Extended Markup Language (VXML). XML wasn't specifically designed for cell phone use or WAP—it's a very important technology that will transform the way the Web is used and built. But it also can be used to build WAP sites and deliver information to cell phones, as you'll see in this chapter. VXML, on the other hand, was specifically designed to give people telephone access to the Internet. Although it can be used with any telephone, it will find its use primarily with cell phones. It allows people to design Web pages that deliver information through voice rather than text and pictures. And it allows people to interact with the Web not by clicking or pushing buttons, but instead by speaking into their telephone.

Chapter 19, "How i-mode Works," covers a wireless Internet technology primarily used in Japan right now, but which will most likely point the way toward how all of us will use the Internet on our cell phones. i-mode has become something of a national obsession in Japan, doing everything from allowing people to send instant text messages to one another to getting daily horoscopes on their cell phones. And it's an always-on technology, so information is delivered without your having to connect—e-mail, for example, could show up on your phone by itself.

Chapter 20, "PCs and Wireless Technology," covers how wireless technologies are commonly used in computers. You'll learn, for example, how wireless mice and keyboards work and how infrared technology allows computers to print without wires.

Finally, Chapter 21, "How Wireless Palmtops Work," covers palmtop computers, sometimes called PDAs, such as the Palm and PocketPC devices. It explains how these devices connect to the Internet and send and receive data and e-mail, and how they are able to beam data to one another using their infrared ports.

CHAPTER 16 Understanding the Internet

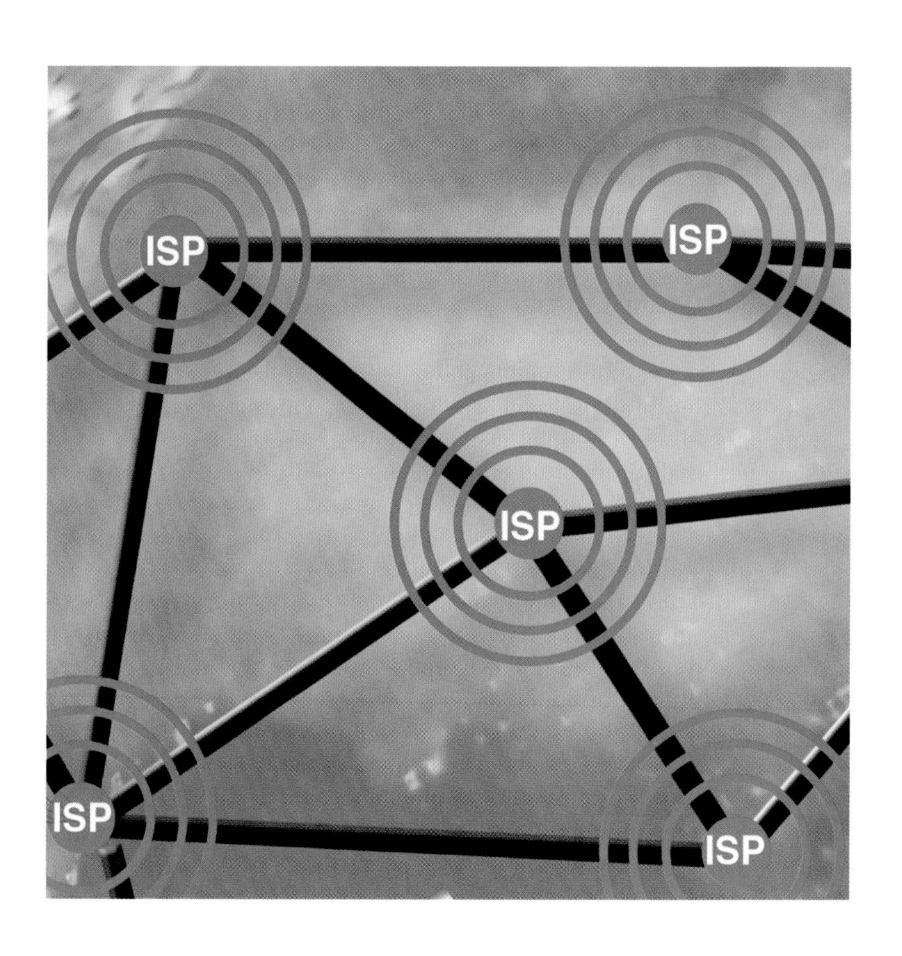

THE next big thing in cell phones—and in personal digital assistants (PDAs) such as the Palm—is accessing the Internet. No longer do you need to be tethered to a computer or laptop to do things such as browse the Web and send and receive e-mail. You can do all that, right now, from your cell phone and PDA.

As you'll see in the rest of the chapters in this section of the book, a variety of technologies enable you to hop onto the Internet with a cell phone or PDA. But no matter what those technologies are, or how they work, they still need to follow the basic way that the Internet works.

Before you can understand how cell phones access the Internet, you first must understand how the Internet itself works. That's what you'll see in the illustrations in this chapter.

Most important to understand about the Internet is that it's a network of networks. These networks can exchange data because all computers on the Internet follow the same basic rules for communicating, with what are called the TCP/IP protocols, which stand for Transmission Control Protocol/Internet Protocol. The Transmission Control Protocol breaks up data that is to be sent across the Internet into small packets, and then reassembles those packets on the computer that receives them. The Internet Protocol handles the job of making sure all the packets get to their proper destination. Doing the job of physically moving all these packets are pieces of hardware called routers.

The Internet is called a *packet-switched network* because the packets are handled this way. When you communicate with someone on a computer, you don't have a single, direct, dedicated connection to that computer or person. Many other people can use the same lines that you are using. The normal telephone, by way of contrast, isn't a packet-switched network—it's a circuit-switched network. When you make a connection with someone, that connection is dedicated only to you and that person, even if neither of you happens to be talking at that point.

An important concept to understand about the Internet, the Web, and e-mail is called client/server. Clients—your Web browser or e-mail software, for example—run on your own computer and request information from a computer on the Internet, known as a server. So, when you browse the Web, your client browser software requests a Web page from a server and then displays it to you. And when you receive e-mail, your client e-mail software requests your e-mail from an Internet mail server (called a POP3) server, and then displays it to you.

How the Internet Works

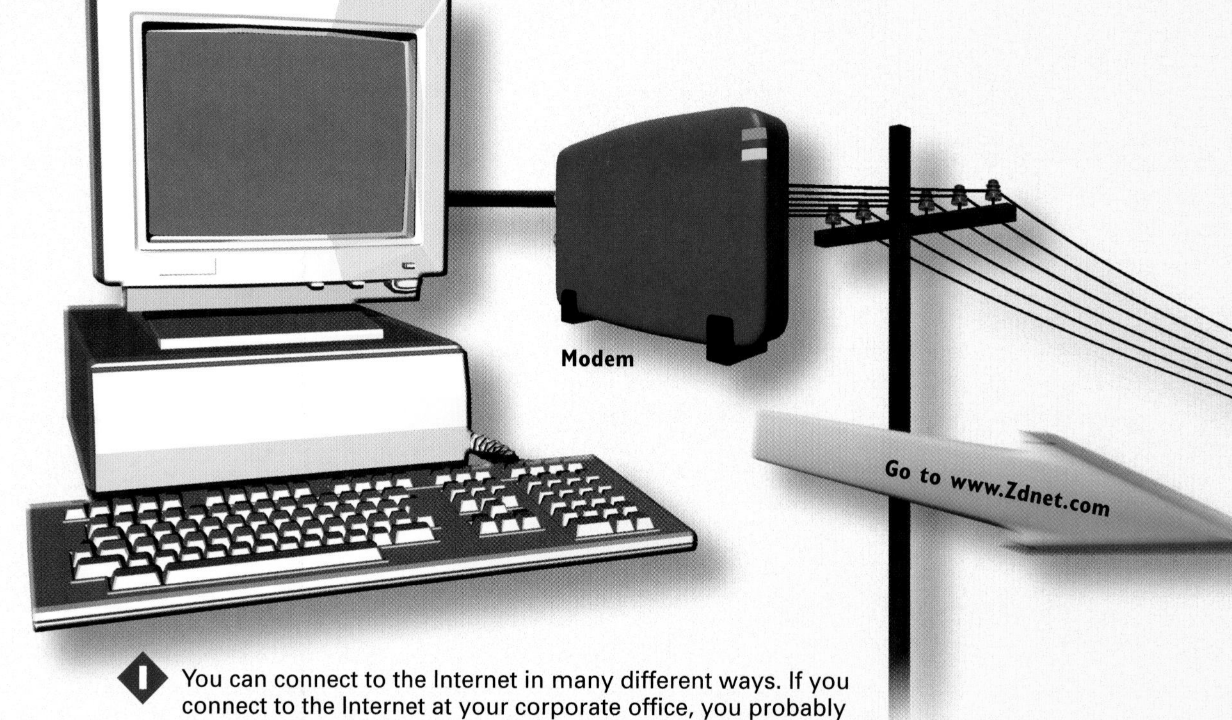

1 You can connect to the Internet in many different ways. If you connect to the Internet at your corporate office, you probably connect through your local area network. If you connect by dialing in at home or by using a device such as a cable modem, you connect through an Internet service provider (ISP), which charges you a monthly fee for the connection. The ISP to which you connect is its own network. When you connect to the Internet, your computer uses a series of protocols called TCP/IP for communicating. The protocols allow computers and networks to talk to one another.

Wireless Tidbit

There's a lot more to learn about the Internet than you'll find in this chapter alone. If you want to see how every aspect of the world's largest and most sophisticated network works, get a copy of *How the Internet Works*. You'll notice that its author might be familiar—I wrote that book as well.

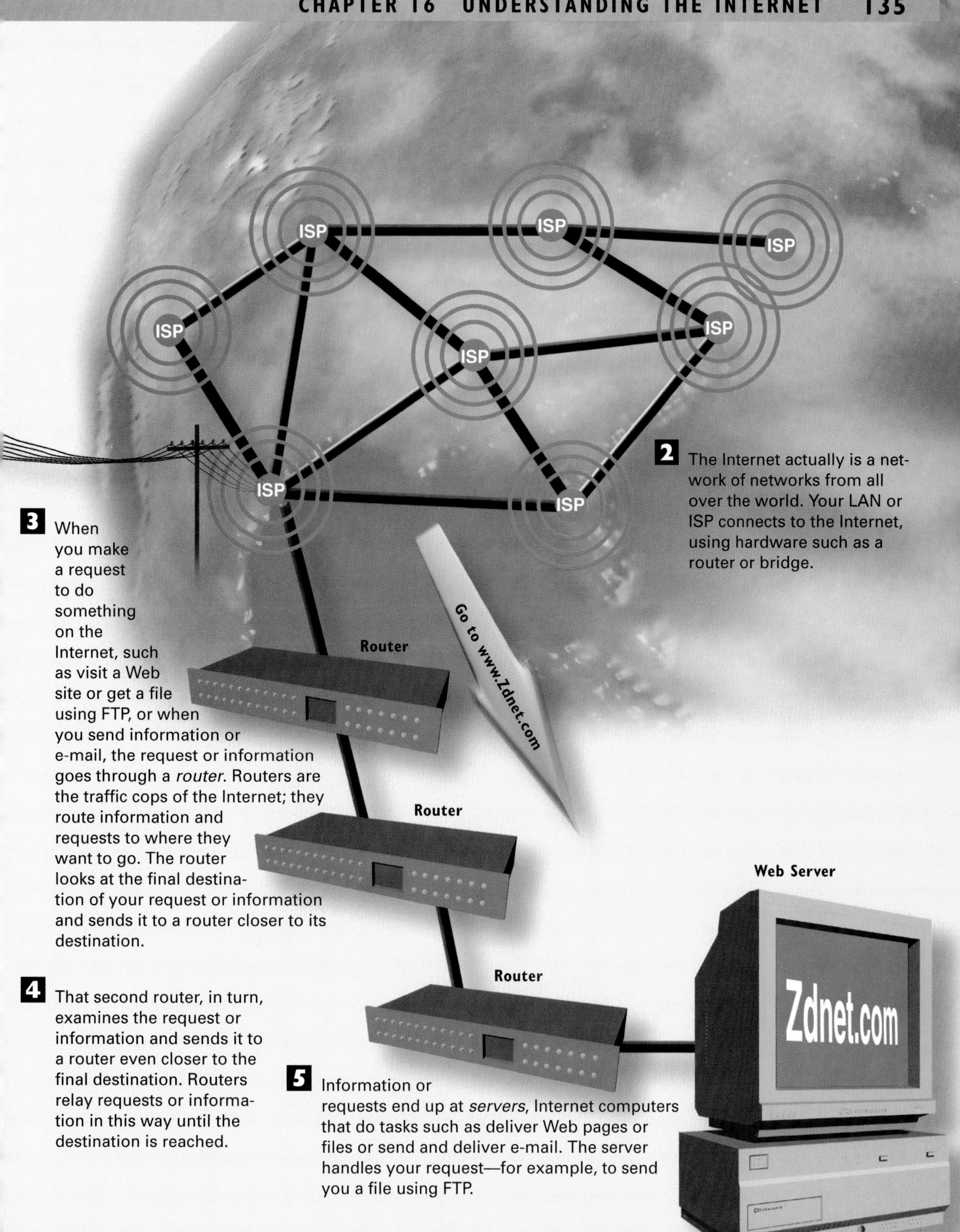

2 The Internet actually is a network of networks from all over the world. Your LAN or ISP connects to the Internet, using hardware such as a router or bridge.

3 When you make a request to do something on the Internet, such as visit a Web site or get a file using FTP, or when you send information or e-mail, the request or information goes through a *router*. Routers are the traffic cops of the Internet; they route information and requests to where they want to go. The router looks at the final destination of your request or information and sends it to a router closer to its destination.

4 That second router, in turn, examines the request or information and sends it to a router even closer to the final destination. Routers relay requests or information in this way until the destination is reached.

5 Information or requests end up at *servers*, Internet computers that do tasks such as deliver Web pages or files or send and deliver e-mail. The server handles your request—for example, to send you a file using FTP.

How TCP/IP Works

1 The Internet is a packet-switched network, which means that when you send information across the Internet from your computer to another computer, the data is broken into small packets. A series of switches called routers send each packet across the Net individually. After all the packets arrive at the receiving computer, they are recombined into their original, unified form. Two protocols do the work of breaking the data into packets, routing the packets across the Internet, and then recombining them on the other end: The Internet Protocol (IP), which routes the data, and the Transmission Control Protocol (TCP), which breaks the data into packets and recombines them on the computer that receives the information.

2 For many reasons, including hardware limitations, data sent across the Internet must be broken up into packets of fewer than 1,500 characters each. Each packet is given a header that contains a variety of information, such as the order in which the packets should be assembled with other related packets. As TCP creates each packet, it also calculates and adds to the header a *checksum*, which is a number that TCP uses on the receiving end to determine whether any errors have been introduced into the packet during transmission. The checksum is based on the precise amount of data in the packet.

TCP

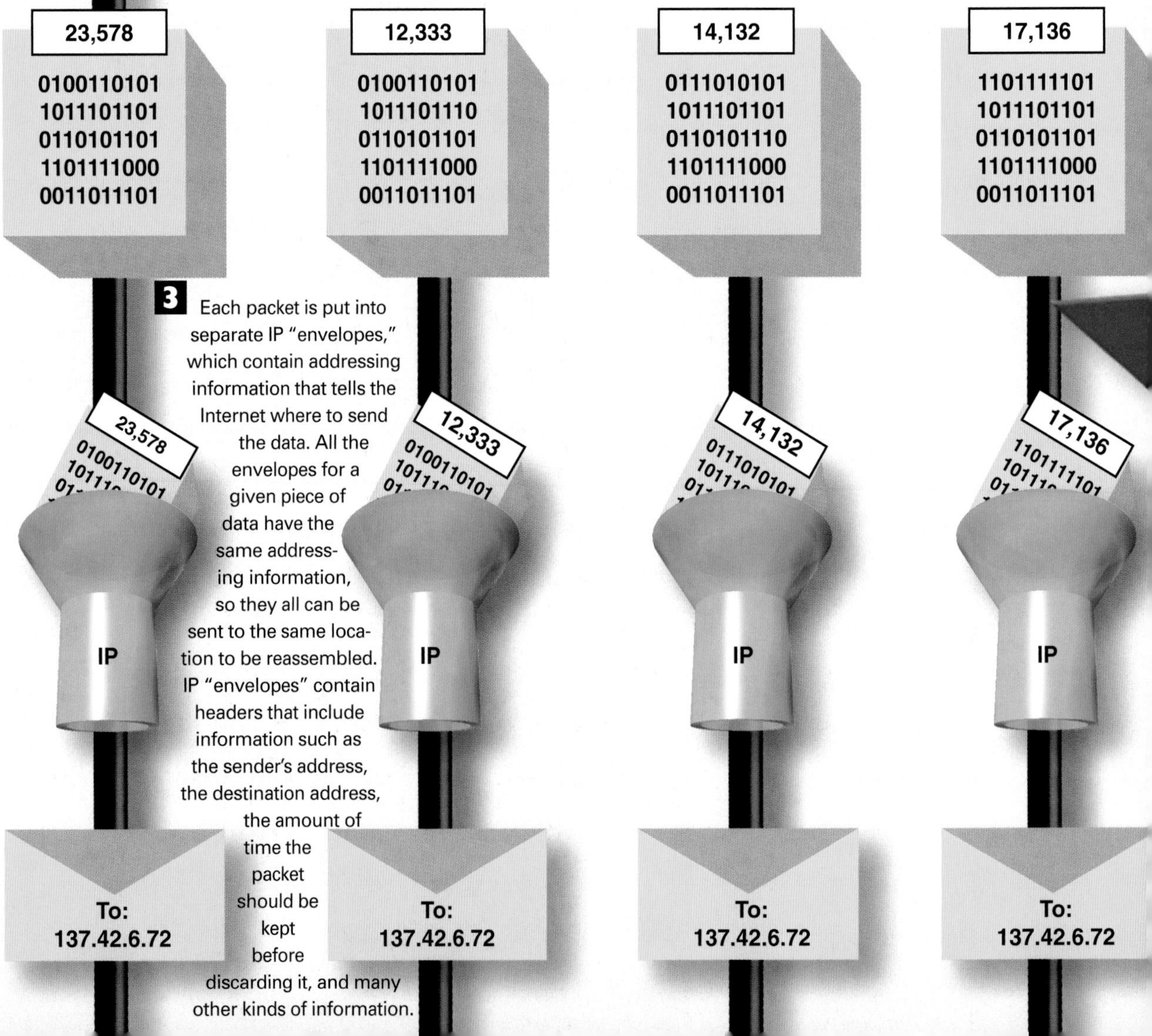

23,578
0100110101
1011101101
0110101101
1101111000
0011011101

12,333
0100110101
1011101110
0110101101
1101111000
0011011101

14,132
0111010101
1011101101
0110101110
1101111000
0011011101

17,136
1101111101
1011101101
0110101101
1101111000
0011011101

3 Each packet is put into separate IP "envelopes," which contain addressing information that tells the Internet where to send the data. All the envelopes for a given piece of data have the same addressing information, so they all can be sent to the same location to be reassembled. IP "envelopes" contain headers that include information such as the sender's address, the destination address, the amount of time the packet should be kept before discarding it, and many other kinds of information.

IP IP IP IP

To: 137.42.6.72

To: 137.42.6.72

To: 137.42.6.72

To: 137.42.6.72

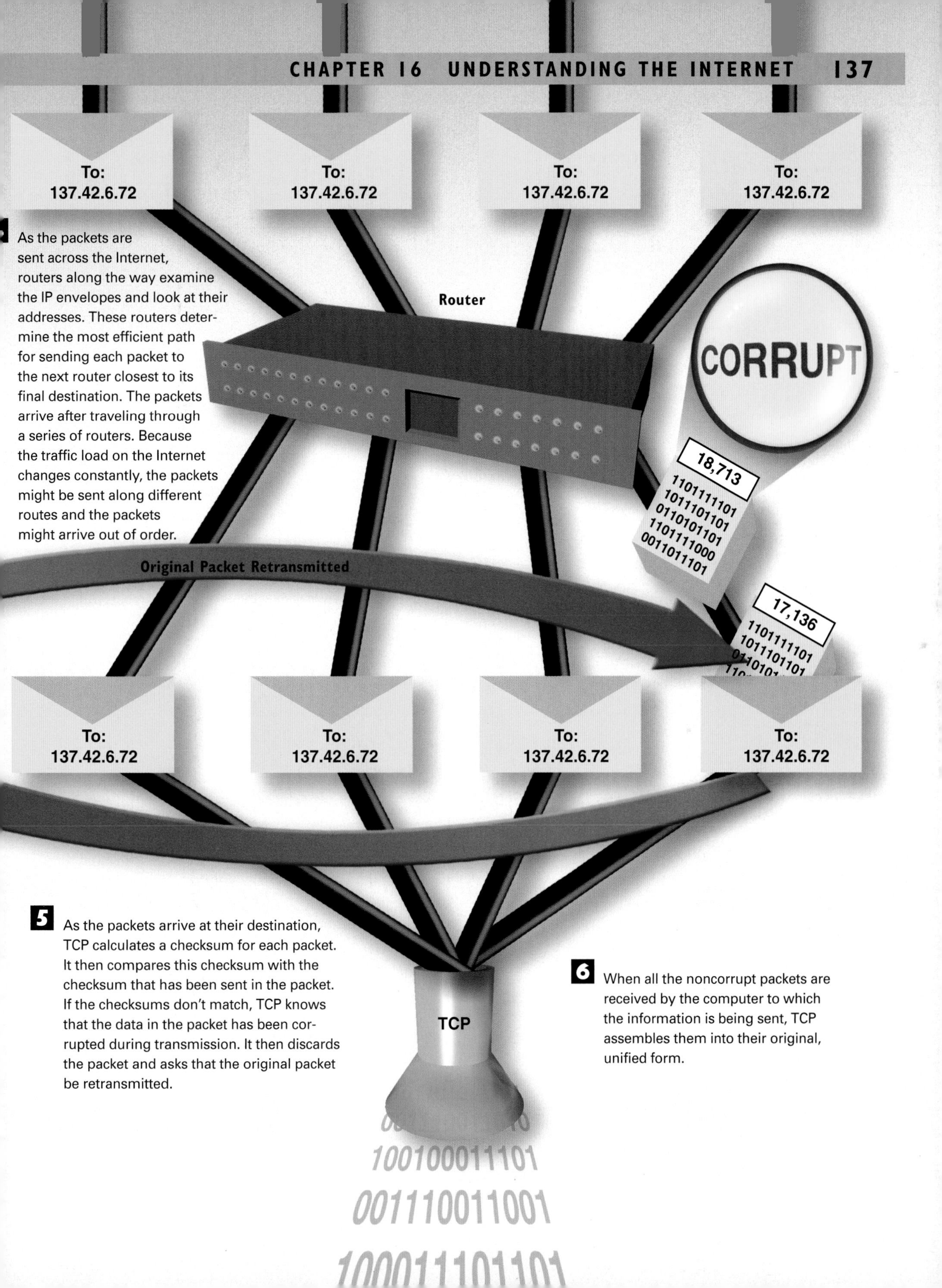

As the packets are sent across the Internet, routers along the way examine the IP envelopes and look at their addresses. These routers determine the most efficient path for sending each packet to the next router closest to its final destination. The packets arrive after traveling through a series of routers. Because the traffic load on the Internet changes constantly, the packets might be sent along different routes and the packets might arrive out of order.

5 As the packets arrive at their destination, TCP calculates a checksum for each packet. It then compares this checksum with the checksum that has been sent in the packet. If the checksums don't match, TCP knows that the data in the packet has been corrupted during transmission. It then discards the packet and asks that the original packet be retransmitted.

6 When all the noncorrupt packets are received by the computer to which the information is being sent, TCP assembles them into their original, unified form.

How the World Wide Web Works

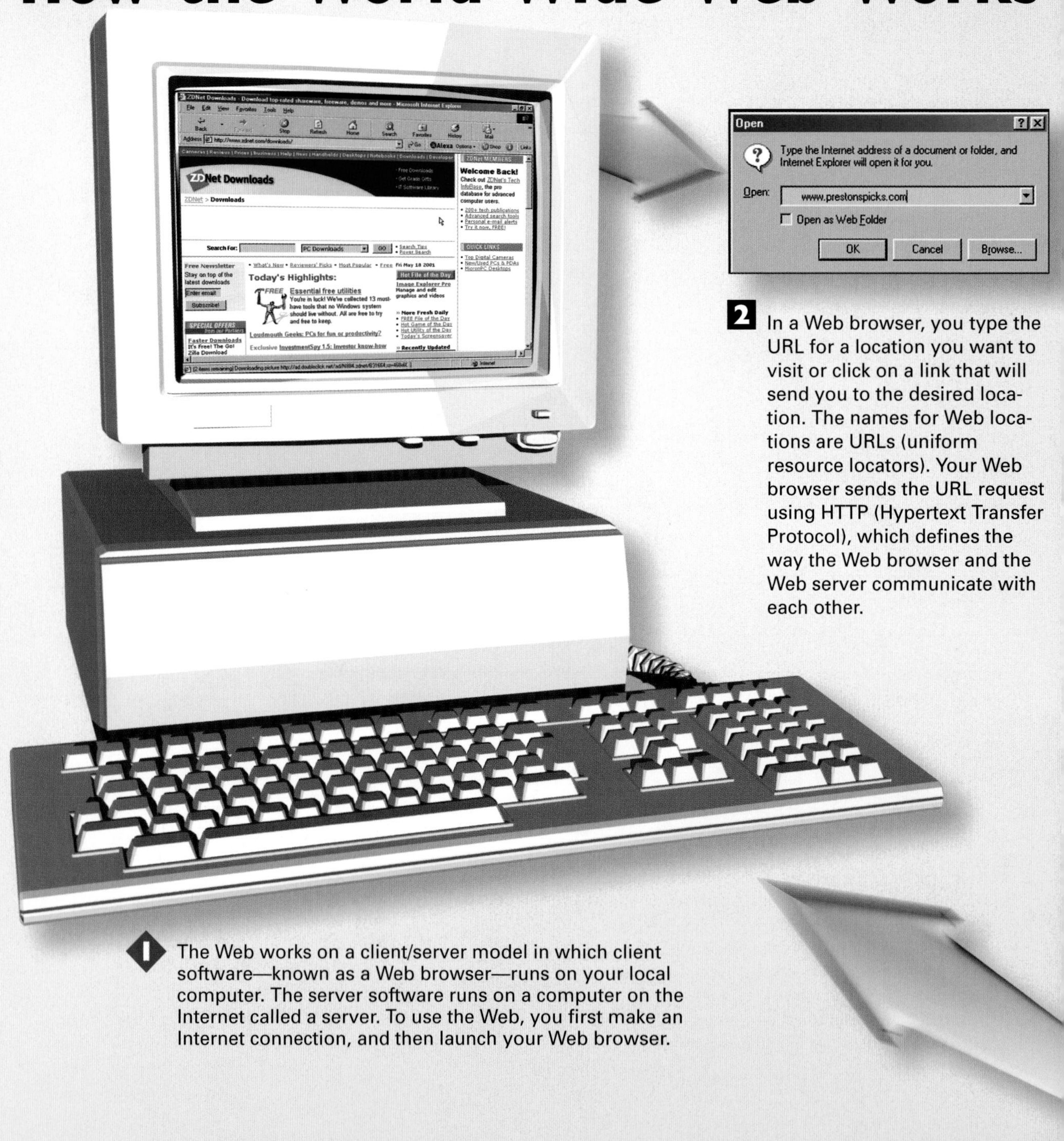

1 The Web works on a client/server model in which client software—known as a Web browser—runs on your local computer. The server software runs on a computer on the Internet called a server. To use the Web, you first make an Internet connection, and then launch your Web browser.

2 In a Web browser, you type the URL for a location you want to visit or click on a link that will send you to the desired location. The names for Web locations are URLs (uniform resource locators). Your Web browser sends the URL request using HTTP (Hypertext Transfer Protocol), which defines the way the Web browser and the Web server communicate with each other.

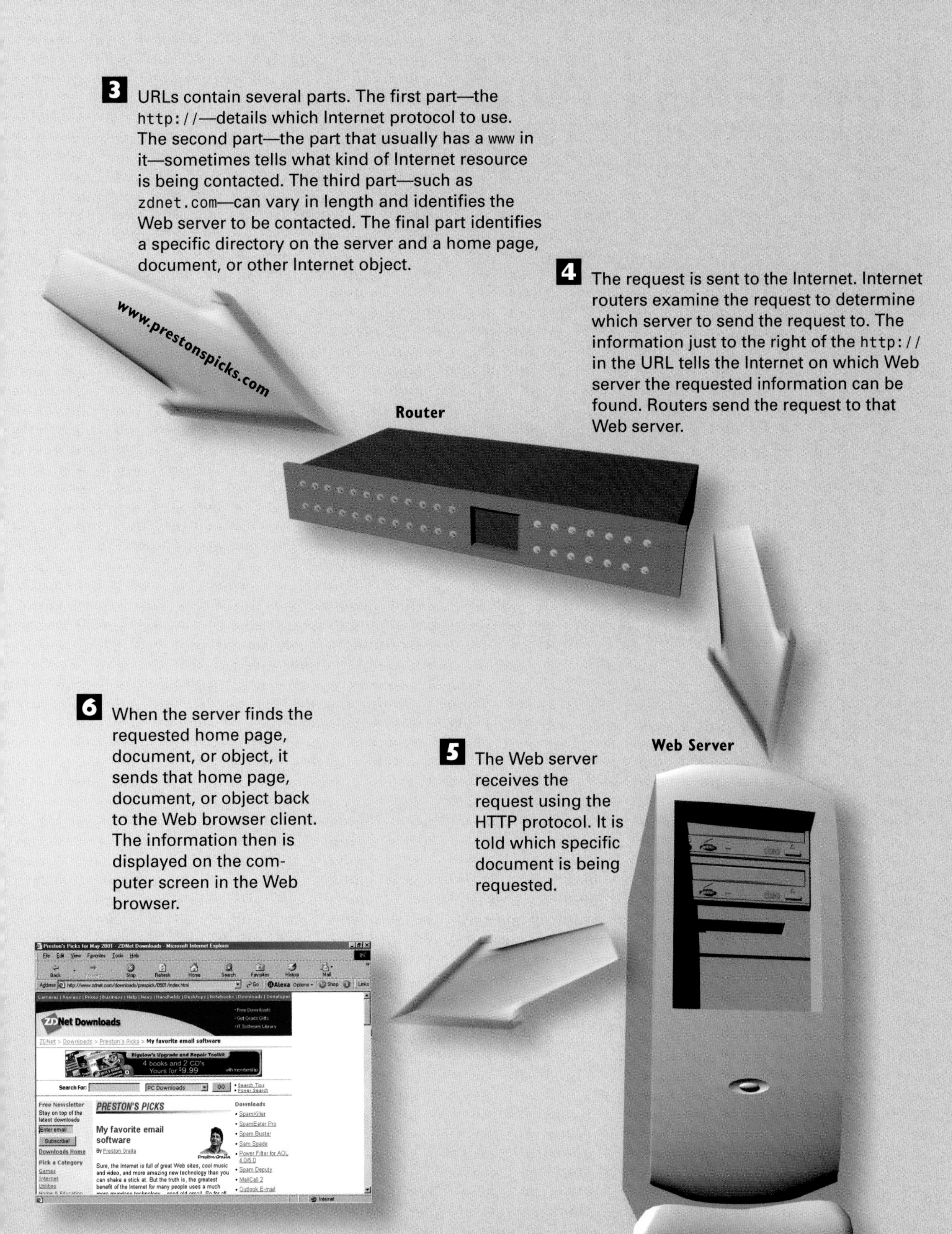

3 URLs contain several parts. The first part—the `http://`—details which Internet protocol to use. The second part—the part that usually has a www in it—sometimes tells what kind of Internet resource is being contacted. The third part—such as `zdnet.com`—can vary in length and identifies the Web server to be contacted. The final part identifies a specific directory on the server and a home page, document, or other Internet object.

4 The request is sent to the Internet. Internet routers examine the request to determine which server to send the request to. The information just to the right of the `http://` in the URL tells the Internet on which Web server the requested information can be found. Routers send the request to that Web server.

5 The Web server receives the request using the HTTP protocol. It is told which specific document is being requested.

6 When the server finds the requested home page, document, or object, it sends that home page, document, or object back to the Web browser client. The information then is displayed on the computer screen in the Web browser.

How E-Mail Works

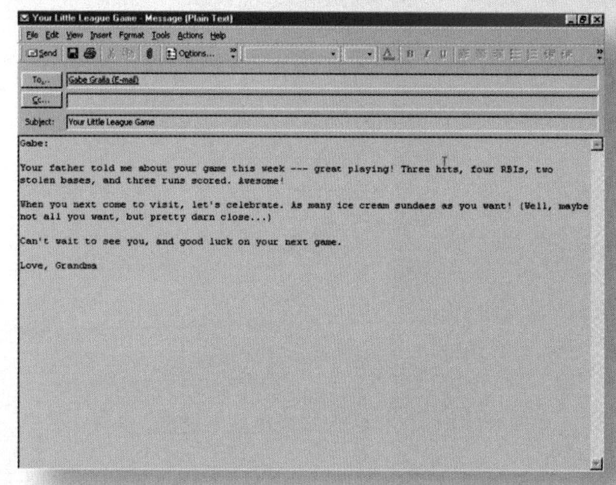

1 You start by creating an e-mail message in your e-mail program and addressing it to whom you want it sent.

2 The e-mail, like all information sent over the Internet, is sent as a stream of packets using the Internet's TCP/IP protocol. Each packet bears the address of the destination. The address it bears is the Internet address—a series of numbers, such as 123.74.78.9—instead of the written address, such as `gabe@gralla.com`.

To: 123.74.78.9.

6 When the intended recipient wants to read e-mail, he logs into the POP3 server using software such as Microsoft Outlook. He then can retrieve all the mail waiting for him. If he wants, he can leave the mail on the server or can have it deleted after he reads it.

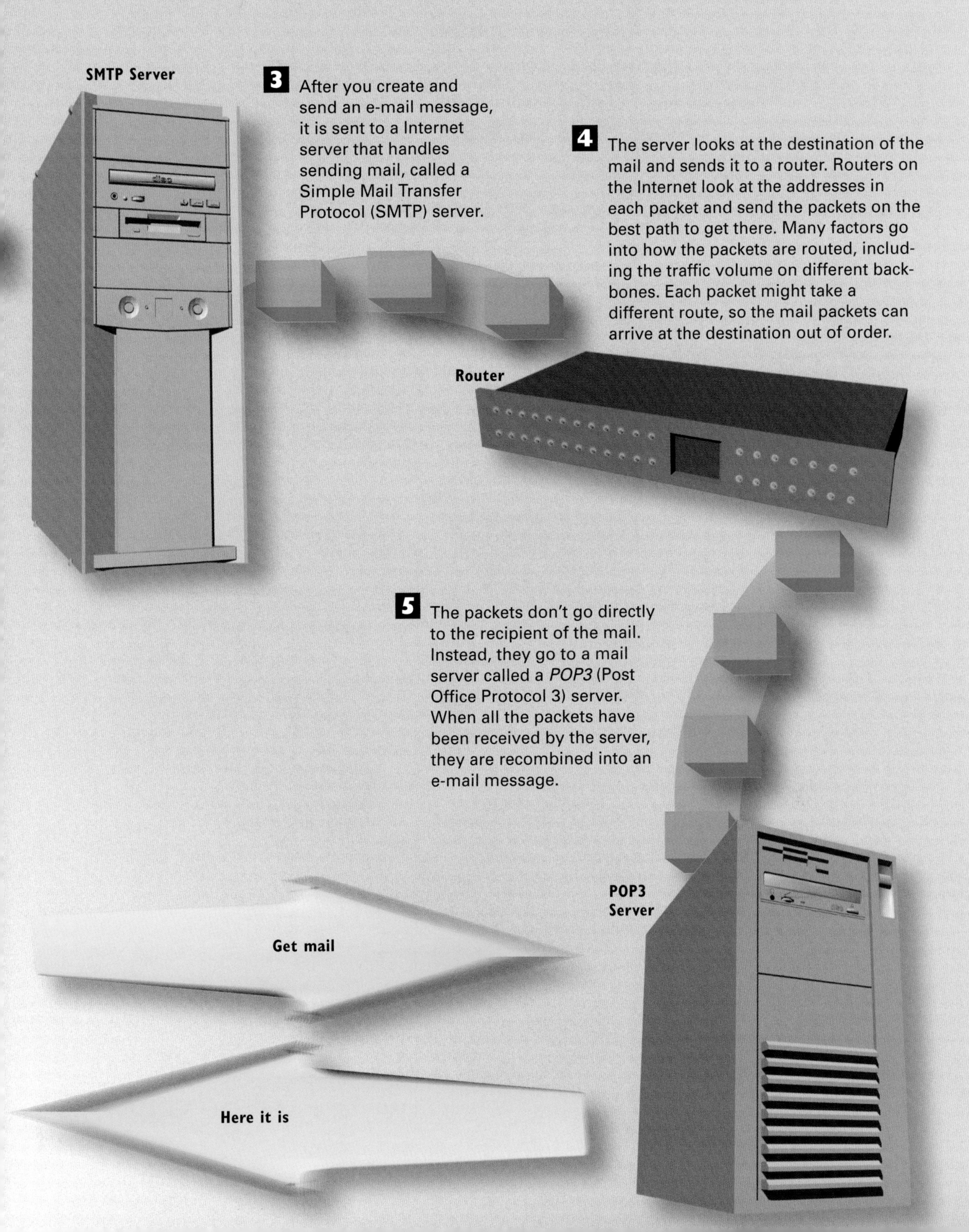

3 After you create and send an e-mail message, it is sent to a Internet server that handles sending mail, called a Simple Mail Transfer Protocol (SMTP) server.

4 The server looks at the destination of the mail and sends it to a router. Routers on the Internet look at the addresses in each packet and send the packets on the best path to get there. Many factors go into how the packets are routed, including the traffic volume on different backbones. Each packet might take a different route, so the mail packets can arrive at the destination out of order.

5 The packets don't go directly to the recipient of the mail. Instead, they go to a mail server called a *POP3* (Post Office Protocol 3) server. When all the packets have been received by the server, they are recombined into an e-mail message.

CHAPTER

17

How Cell Phones Access the Internet

THE Internet, once the domain solely of computers, now can be accessed by anyone with a cell phone. Increasingly, many of the new cell phones being manufactured include the capability to use the Internet, and many cell phone calling plans include Internet access as part of them.

But it's no easy task to allow a cell phone, with its tiny screen and usually low-speed connection, to access the Internet. To allow cell phones to access the Internet, the *Wireless Access Protocol (WAP)* is used. WAP was specifically designed with low-speed, cellular connections in mind, and recognizing that the device contacting the Internet will have a small window, not a full-size Web browser. WAP protocols are built not just for the present, but for the future as well. So, although WAP works well with today's low-speed cellular Internet connections, it's also designed so that when wireless high-speed 3G connections come into being, cell phones will be able to take full advantage of them.

WAP actually is a full suite of protocols and services, not a single one. For example, it includes the WAP Transaction Protocol (WTP), which is in a way the equivalent of the TCP/IP protocols that form the underlying basis of the Internet. And it also includes the Wireless Transport Layer Security (WTLS), which allows for the sending and receiving of encrypted information so that you can feel secure in shopping and banking when you use it.

Two of the most important components of WAP are WML, the Wireless Markup Language; and WMLScript, a scripting language that allows for interaction between the cell phone and the Internet.

WML is based on HTML, the language that is used to build Web pages. WML is made up of commands specifically designed to display text—and even graphics—on a small cell phone screen. Designers create sites with WML; when your cell phone contacts a site, it's actually downloading WML documents to your cell phone. A *microbrowser* in your cell phone then displays those documents, in the same way that a browser displays Web pages on a computer screen.

WMLScript is similar to the Web's JavaScript scripting language and, like JavaScript, it adds interactivity to pages. Especially important is that it can have the microbrowser that otherwise would have to be done by contacting a Web site. This is important, because cell phones usually have low-speed connections to the Internet, so WMLScript can speed up the delivery of information. And it also can help keep cell phone costs down.

Another important component of WAP is the WAP gateway. This is a special server that translates requests and information between the TCP/IP protocols of the Web and the WAP protocols of cell phones. The gateway also can reformat Web pages so that they display better on cell phones, although the truth is, unless a Web site is built specifically for WAP and cell phones, it never looks very good.

How Web Pages Are Delivered with WAP

1 The primary way in which cellular telephones access the Internet is through a protocol called the Wireless Access Protocol (WAP) and its associated markup language, the Wireless Markup Language (WML). To browse the Web with a cell phone, the phone must have what's called a *microbrowser*—the capability to use WAP, interpret WML, and display Web pages on a small screen. Not all cell phones have microbrowsers, and the microbrowsers on different phones sometimes have different capabilities.

2 The cell phone connects to a cell and requests to visit a Web page.

7 The WML page is sent back through the landline to a base station. The base station sends the page to your cellular telephone.

8 You now can read the page on your cellular telephone—it's been specially formatted for its display. However, cellular phones have difficulty handling graphics, so not all Web pages will display properly, even after they've been reformatted to WML. And even pages written in WML can have problems displaying, because not all microbrowsers can handle all WML commands.

Wireless Tidbit

Wireless palmtop computers, such as the Palm, can use WAP to display Web pages. However, to do so, they need to download and install special WAP software.

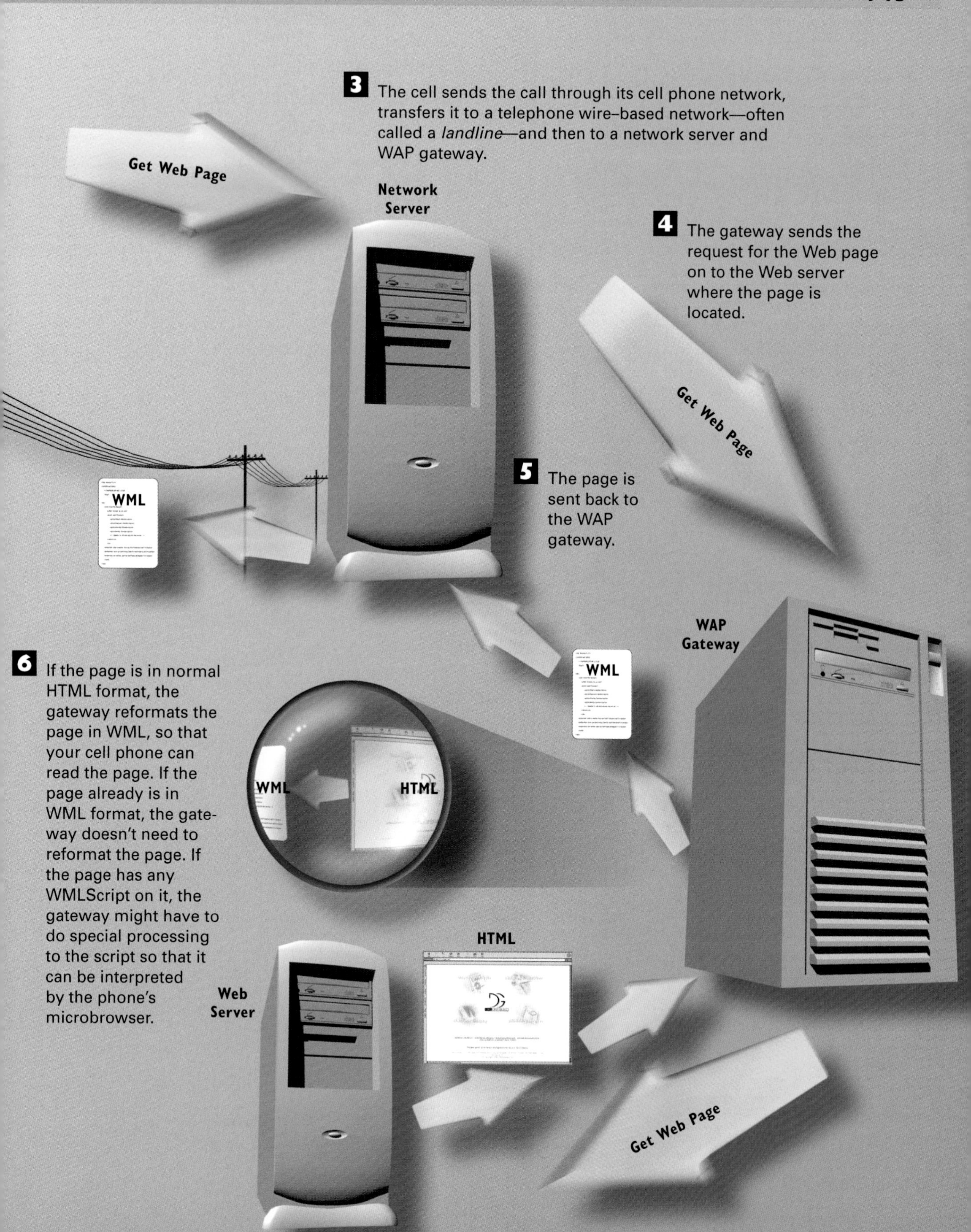

3 The cell sends the call through its cell phone network, transfers it to a telephone wire–based network—often called a *landline*—and then to a network server and WAP gateway.

4 The gateway sends the request for the Web page on to the Web server where the page is located.

5 The page is sent back to the WAP gateway.

6 If the page is in normal HTML format, the gateway reformats the page in WML, so that your cell phone can read the page. If the page already is in WML format, the gateway doesn't need to reformat the page. If the page has any WMLScript on it, the gateway might have to do special processing to the script so that it can be interpreted by the phone's microbrowser.

How the Wireless Markup Language (WML) Works

1. To create pages that can be displayed on a cell phone's microbrowser, WML tags are added to the text to be displayed. All WML files begin and end with WML tags. WML allows for emphasis to be added to the text, with tags that do things such as add boldface, italic, and underlining. WML is based on the Web's HTML language and is designed specifically for cell phone displays. WML doesn't have nearly as many controls over the appearance of text as does HTML—it can't do things such as specify a particular typeface. And in any event, all microbrowsers are different, and many microbrowsers ignore the tags.

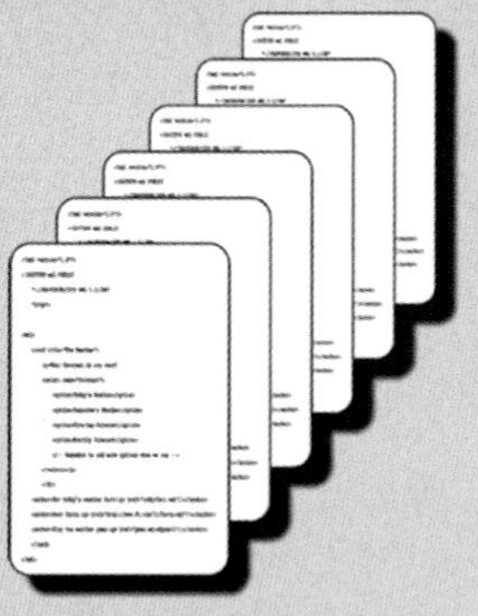

WML Deck

2. WML documents are organized into *cards*. One card at a time is displayed on the microbrowser. This is unlike HTML, which allows individual HTML documents to be very long and be scrolled. There can be more than one card in a WML document. These cards are all related and display related information. Taken together, all the cards in a WML document are called a *deck*.

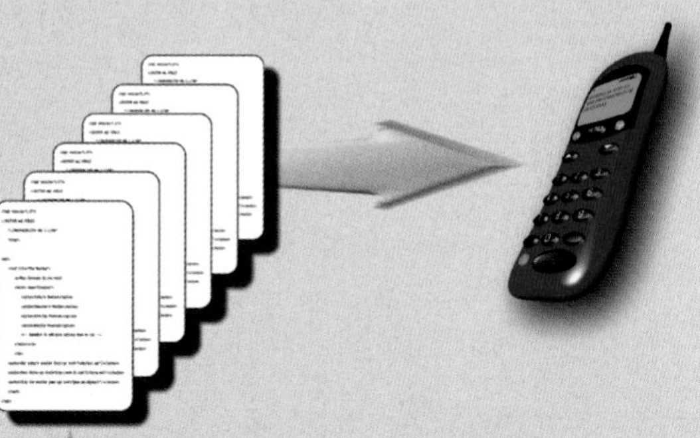

Server

3. When a microbrowser contacts a WML-based page, it downloads all the cards in a deck at one time. Doing this means that the microbrowser won't have to go back to the site every time a new card in the deck needs to be downloaded—the card will be right on the cell phone. WML page designers must be careful when designing their documents to be sure only to have cards downloaded that are truly required by the cell phone; otherwise, the downloading will take a great deal of time.

Wireless Tidbit

WML designers often like to put comments to themselves inside a WML document—so that, for example, they'll remember why they put in a particular WML command. They surround their comments with special tags so that the comments won't be displayed on the microbrowser. The comments aren't even delivered to the cell phone—the WAP gateway looks for them and automatically deletes them before sending the WML document to the cell phone. It does this to cut down on the amount of data transmitted, so that the document can be sent more quickly, and users charged less money, if they happen to pay according to the amount of data they download to their cell phone.

4 When someone contacts a link on a card, if the linked card is on the cell phone, the new card will be displayed on the microbrowser.

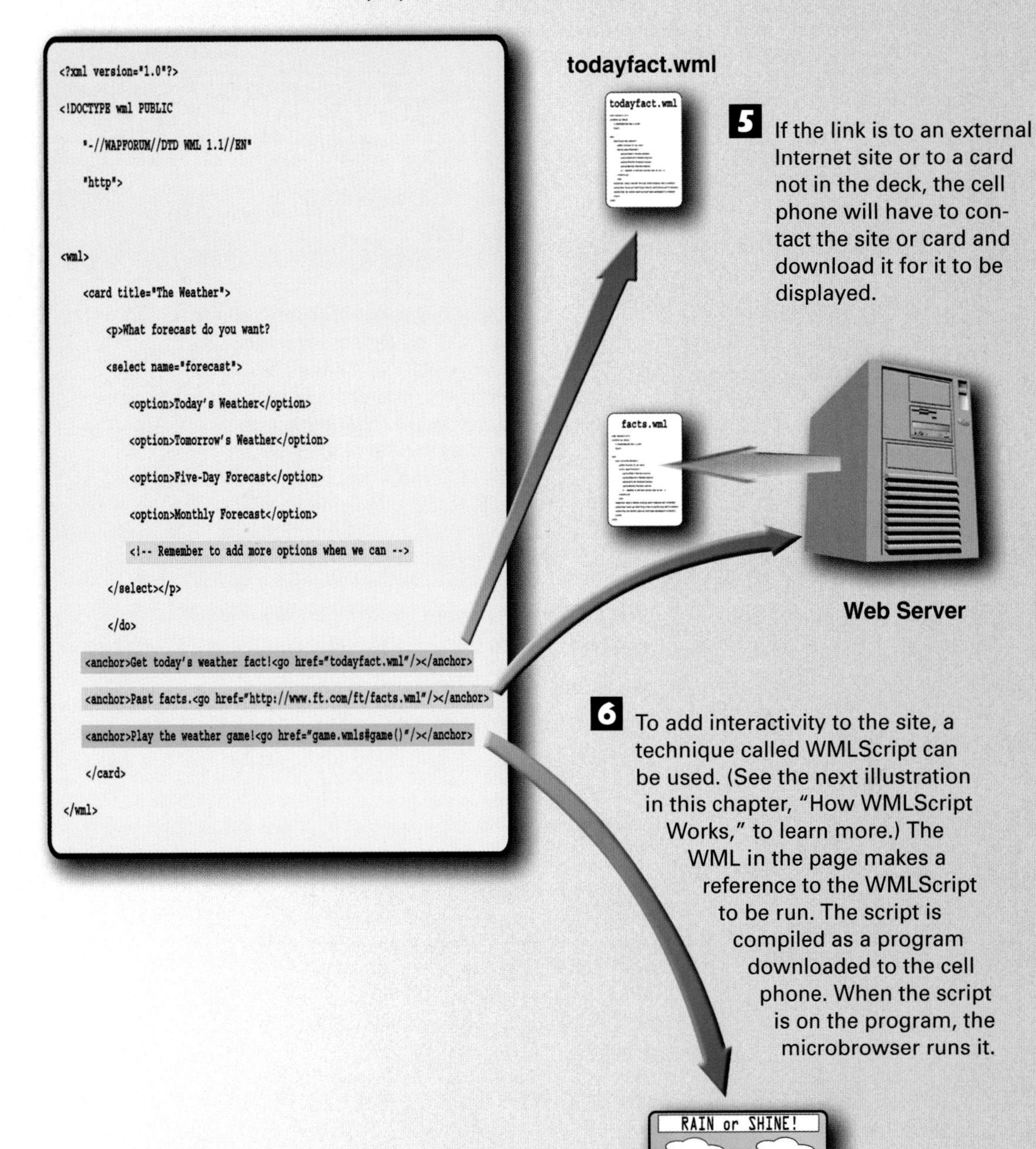
<?xml version="1.0"?>
<!DOCTYPE wml PUBLIC
"-//WAPFORUM//DTD WML 1.1//EN"
"http">
<wml>
<card title="The Weather">
<p>What forecast do you want?
<select name="forecast">
<option>Today's Weather</option>
<option>Tomorrow's Weather</option>
<option>Five-Day Forecast</option>
<option>Monthly Forecast</option>
<!-- Remember to add more options when we can -->
</select></p>
</do>
<anchor>Get today's weather fact!<go href="todayfact.wml"/></anchor>
<anchor>Past facts.<go href="http://www.ft.com/ft/facts.wml"/></anchor>
<anchor>Play the weather game!<go href="game.wmls#game()"/></anchor>
</card>
</wml>
todayfact.wml
todayfact.wml
facts.wml
Web Server
RAIN or SHINE!
Guess today's high! 55° 60° 65° 70°

5 If the link is to an external Internet site or to a card not in the deck, the cell phone will have to contact the site or card and download it for it to be displayed.

6 To add interactivity to the site, a technique called WMLScript can be used. (See the next illustration in this chapter, "How WMLScript Works," to learn more.) The WML in the page makes a reference to the WMLScript to be run. The script is compiled as a program downloaded to the cell phone. When the script is on the program, the microbrowser runs it.

How WMLScript Works

1. WMLScript is a scripting language loosely based on the JavaScript scripting language used to deliver information to Web browsers, although it's a much simpler language. It's used to add interactivity to WML pages. First, a programmer writes a script using the language.

```
/*
 * A game of battleships.
 */
extern function init (  )
```

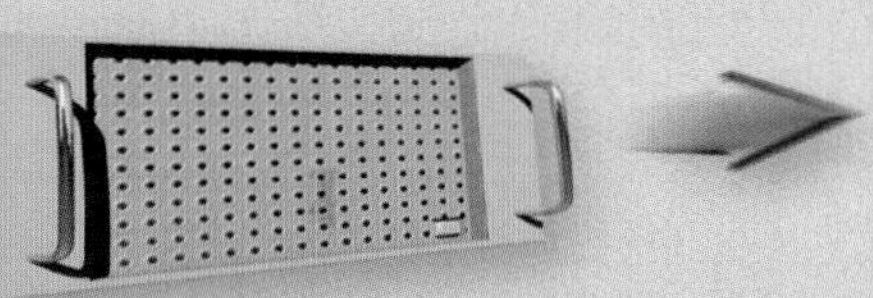

2. Next, the script is compiled into something called *bytecode*, which is, in essence, a series of instructions that a basic computer can understand. That bytecode ultimately will be run inside a *WMLScript interpreter*, also called a *WMLScript virtual machine*, inside the cell phone's microbrowser. Microbrowsers include a WMLScript interpreter, so no extra software needs to be downloaded to run WMLScript. If a cell phone can access the Internet with a microbrowser, it can run WMLScript.

3. The bytecode is put onto a Web server.

Wireless Tidbit

To do its work, WMLScript can use several "libraries" found in the microbrowser. These libraries can do many things, such as arithmetic and display dialog boxes to the user to ask whether he wants to go ahead with a certain task. To use these libraries, specific syntax is used when writing the WMLScript.

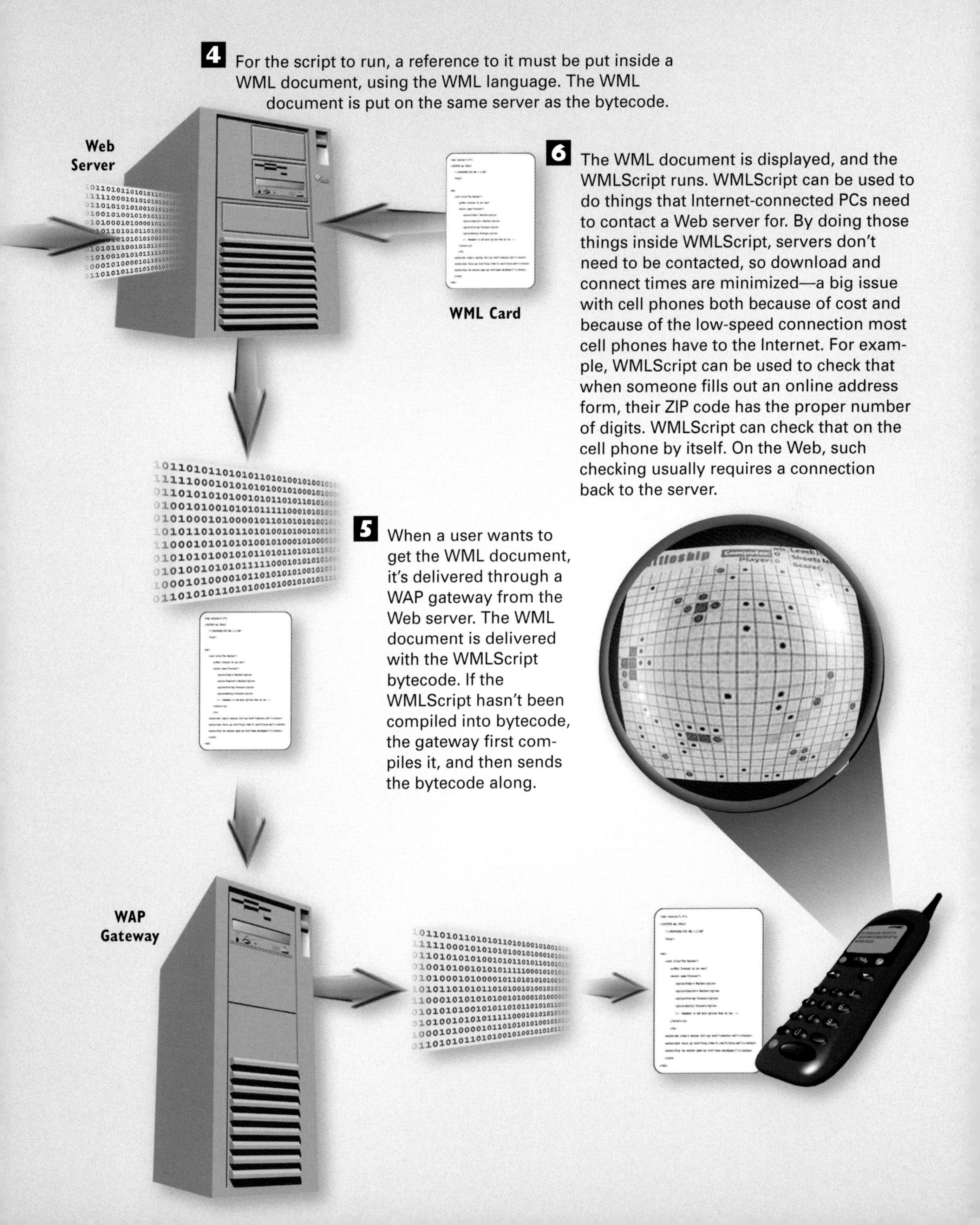

4 For the script to run, a reference to it must be put inside a WML document, using the WML language. The WML document is put on the same server as the bytecode.

5 When a user wants to get the WML document, it's delivered through a WAP gateway from the Web server. The WML document is delivered with the WMLScript bytecode. If the WMLScript hasn't been compiled into bytecode, the gateway first compiles it, and then sends the bytecode along.

6 The WML document is displayed, and the WMLScript runs. WMLScript can be used to do things that Internet-connected PCs need to contact a Web server for. By doing those things inside WMLScript, servers don't need to be contacted, so download and connect times are minimized—a big issue with cell phones both because of cost and because of the low-speed connection most cell phones have to the Internet. For example, WMLScript can be used to check that when someone fills out an online address form, their ZIP code has the proper number of digits. WMLScript can check that on the cell phone by itself. On the Web, such checking usually requires a connection back to the server.

How Cell Phones Send and Receive E-Mail

2 When you want to get your e-r
you connect to the Internet as
normally do with your cell pho
Your cell phone connects throu
a WAP gateway.

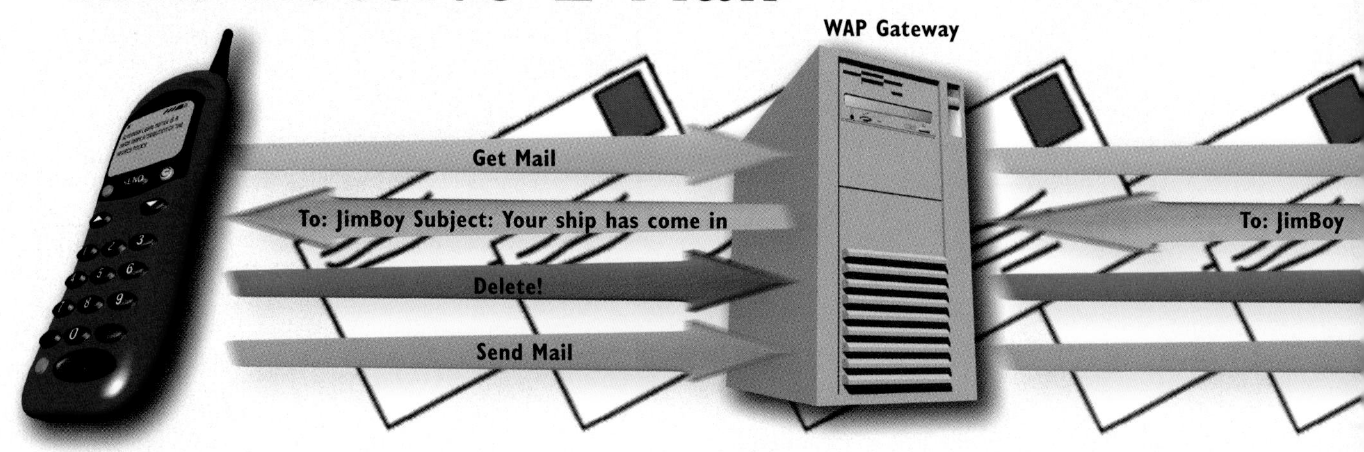

5 After you read the mail, you can delete it from the server or leave it on the server so that you can later have it on your computer, if you contact the mail server through your computer.

3 To make it easier to get your e-mail from a POP3 or IMAP server, you can connect to a special cell phone e-mail gateway and portal. You'll have to configure this gateway and portal with the exact name and location of your POP3 server, such as `pop3.email.net`. This gateway offers special functionality for users of cell phones who contact their computer's e-mail servers, such as only displaying the headers of the messages and not the entire message until you ask for it.

Wireless Tidbit

When you get an Internet-enabled cell phone, it often comes with an e-mail box. If you want to use this inbox and not bother to check your computer-based e-mail account, you won't have to do any special configurations, such as finding out the server name of your POP3 mail server.

6 If you want to respond or send e-mail to someone else, you compose an e-mail message on your cell phone. You again go through the WAP gateway and the e-mail gateway and portal—except to send mail, you don't contact your POP3 or IMAP server because those servers only receive mail, and don't send mail. Instead, you contact an SMTP (Simple Mail Transfer Protocol) server, which sends your e-mail.

Mail Portal

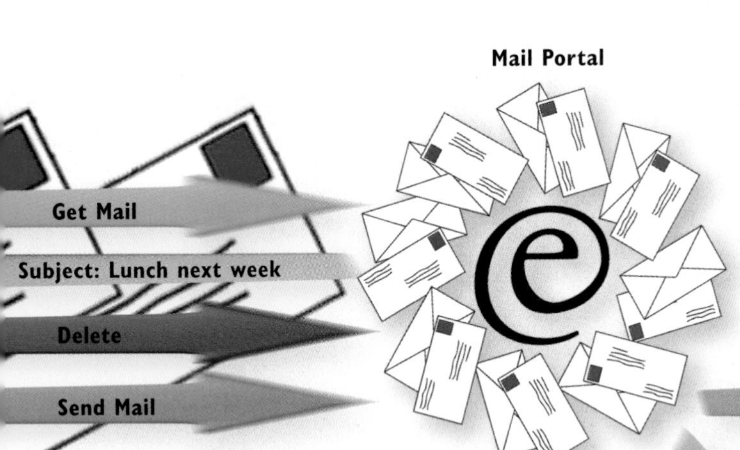

SMTP Server

4 From the gateway and portal, you can check your inbox by logging into the server as you normally do on your computer. The contents of your inbox are sent through the e-mail gateway and portal, to the WAP server, and then to your cell phone. Although there are many ways you can check your e-mail box, typically, you'll download just the headers to your cell phone. You then can select one or several messages to read. After reading a header, if you want to read a message, your cell phone makes the entire circuit again, requesting the mail—from WAP gateway to e-mail gateway and portal to your mail server. You can download the entire message or just a portion of it.

POP3 Server

1 There are many different ways that cell phones can send and receive e-mail. In this illustration, we'll look at how cell phones can send and receive e-mail using the same e-mail box that your computer normally uses. Your computer's e-mail is kept on a special mail server, called a POP3 (Post Office Protocol 3) or IMAP (Internet Message Access Protocol) server. Your e-mail sits on an inbox on a server until you contact it. It doesn't matter to the server what kind of device contacts it—a computer, a cell phone, a personal digital assistant (PDA), or some other device.

CHAPTER

18

How XML and Voice XML Deliver Internet Data

WIRELESS technology is relatively new, and giving wireless devices, such as wireless phones and personal digital assistants (PDAs), access to the Internet is even newer. Because of that, it's still not yet thoroughly clear how Internet information will be delivered to wireless devices. However, many people believe that one of the primary ways will be through an Internet technology called *eXtensible Markup Language (XML)*. This language is an outgrowth of the markup language that forms the underlying basis of the Web, the Hypertext Markup Language, or HTML.

To understand XML, and how it will deliver information to wireless devices, you first must understand a little bit about HTML. HTML instructs a Web browser how to view a page—for example, to display certain text as large and another as small; to display graphics; and so on. HTML *documents* are text files placed on a Web server. When a computer visits the server, the page is downloaded to the computer, and the browser displays the page based on the commands in HTML.

As you learned in Chapter 17, "How Cell Phones Access the Internet," there is an HTML variant called Wireless Markup Language (WML) specifically designed so that when cell phones visit Web pages, the pages will be displayed properly on their small screens.

XML takes a different approach. It's not designed to tell browsers how to display information—in fact, it *can't* tell a browser how to display information, because it doesn't contain those kinds of display commands. Instead, XML marks up the contents of a page and defines what kind of content each different element is. For example, if XML were used to define a book, there would be a set of tags defining the chapter number, another set of tags defining the chapter title, another set of tags defining the chapter text, and so on.

What makes XML so important is that it separates the contents of a page from its display. So, after the content is defined, it can be displayed many different ways by applying different templates. The content on the page never needs to change—one just needs to create or change a template. It's like looking at the exact same weather report in several different newspapers in which the information is the same, but the colors and style of the weather map are different.

This is important for wireless access to the Web, because it means that Web designers can use XML to create a page only once, and then have different templates applied to it so that the page looks one way to a computer, another way to a cell phone, and so on.

An interesting variant of XML, *Voice eXtensible Markup Language (VXML)*, will be used to deliver Internet content to cell phones as well. VXML allows designers to create Web sites that are never viewed—instead, the pages are read to visitors. And visitors interact with the pages simply by speaking into the phone. Considering how annoying it can be to use cell phone keypads, this could be a big step forward for delivering Internet content to cell phones.

How XML Works

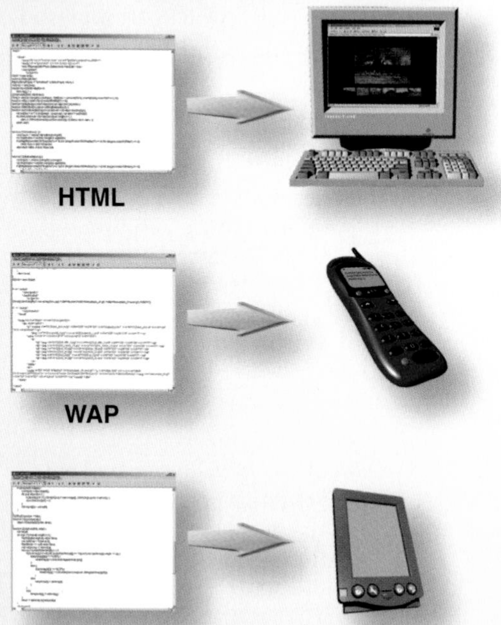

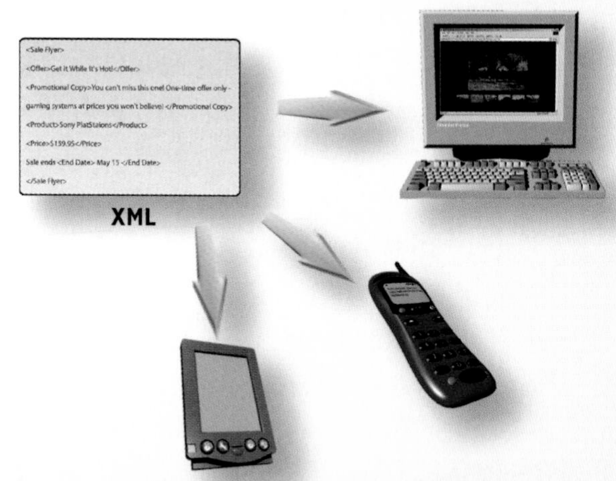

2 With XML, a developer can create the Web site just once. Then, it can be automatically formatted to several different kinds of devices, such as Internet-connected computers and cell phones, using WAP.

1 XML isn't confined to the wireless Web—it's finding widespread use all over the Internet—but it solves a major problem for Web developers. Without XML, if a Web developer wanted to deliver information to Internet-connected computers, to cell phones using WAP, and to Palm devices using Web clipping technology, the developer would have to create and maintain three separate Web sites—an expensive and time-consuming proposition. Because of this, much Web-delivered content to cell phones probably will be built using XML because often it won't make financial sense to build a wireless-only Web site.

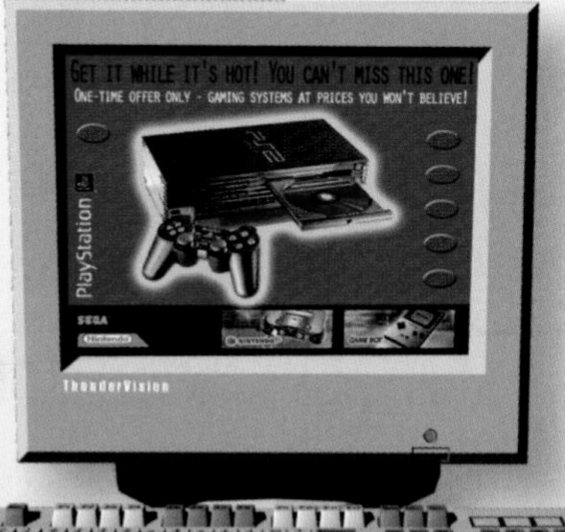

Get it while it's hot! You can't

4 When XML content is posted on a Web site, different designs must be applied to that content so that it can be viewed by devices connecting to it—for example, cell phones. *eXtensible Style Language Transformations (XSLT)* can be applied to the XML. XSLT can take XML and apply different designs to it, or change it to other forms of XML. For example, it can take the XML and turn it into a WAP page that can be viewed by a cell phone, and take the same XML and turn it into an HTML document with a different design.

```
<Sale Flyer>
<Offer>Get It While It's Hot!</Offer>
<Promotional Copy>You can't miss this one! One-time offer only -
gaming systems at prices you won't believe! </Promotional Copy>
<Product>Sony PlatStaions</Product>
<Price>$159.95</Price>
Sale ends <End Date> May 15 </End Date>
</Sale Flyer>
```

XML Document

3 The most important concept to understand about XML is that the language is used only to convey information about content, not about the presentation of the content. So, for example, it doesn't give instructions on what size text should be. But it uses tags to define the kind of content on the page. Then it uses other techniques, as you'll see in the next steps, to display those pages. In that way, a single page can be displayed many different ways, without having to go back and alter the original page—only the designs, which are separate from the content, need to be changed.

5 When a cell phone visits a site built with XML, there must be some way for the site to know that the device is a cell phone and requires WAP, rather than a normal PC. Web-based *Common Gateway Interface (CGI)* scripts can accomplish that by sending queries to the cell phone and listening for their answers.

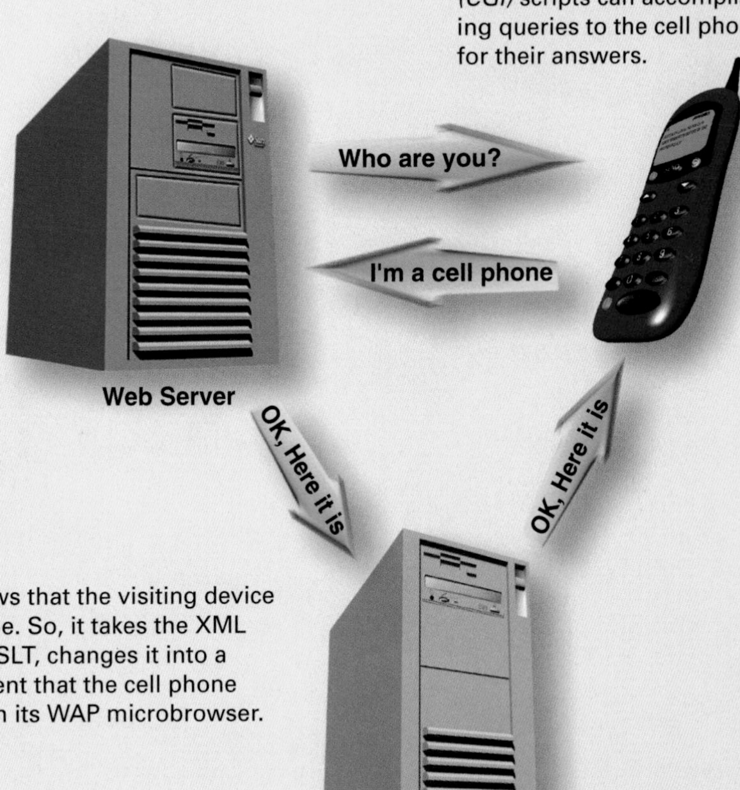
Who are you?
I'm a cell phone
Web Server
OK, Here it is
OK, Here it is
WAP Gateway

6 The site knows that the visiting device is a cell phone. So, it takes the XML and, using XSLT, changes it into a WAP document that the cell phone can view with its WAP microbrowser.

How Voice eXtensible Markup Language (VXML) Works

1 VXML is a variant of XML, with its own special instructions for accepting voice input and delivering information via voice. For someone with a cell phone to get information or order something using VXML, a VXML document must be coded and put onto a Web server.

```
<?xml version="1.0"?>
<vxml version="1.0">
  <menu>
  <prompt>Would you like <enumerate/></prompt>

  <choice next="http://...today.vxml"> Today's weather </choice>
  <choice next="http://...tomorrow.vxml"> Tomorrow's weather</choice>
  <choice next="http://...fiveday.vxml"> The five-day forecast </choice>

  <nomatch>I didn't understand what you said.</nomatch>
  <noinput>Please say something to make a choice.</noinput>
  </menu>
</vxml>
```

2 When someone wants to get information or order something from a VXML site, he connects using his cell phone, just as he would make any other call.

3 The call doesn't go to a Web server or Web site. Instead, it goes to a VXML gateway that handles much of the VXML processes. The gateway contains three primary components: a voice browser, which interprets VXML commands for the telephone; an automated speech recognition (ASR) component that can recognize spoken words and send the information to the VXML document; and a text-to-speech (TTS) component that takes text and turns it into speech.

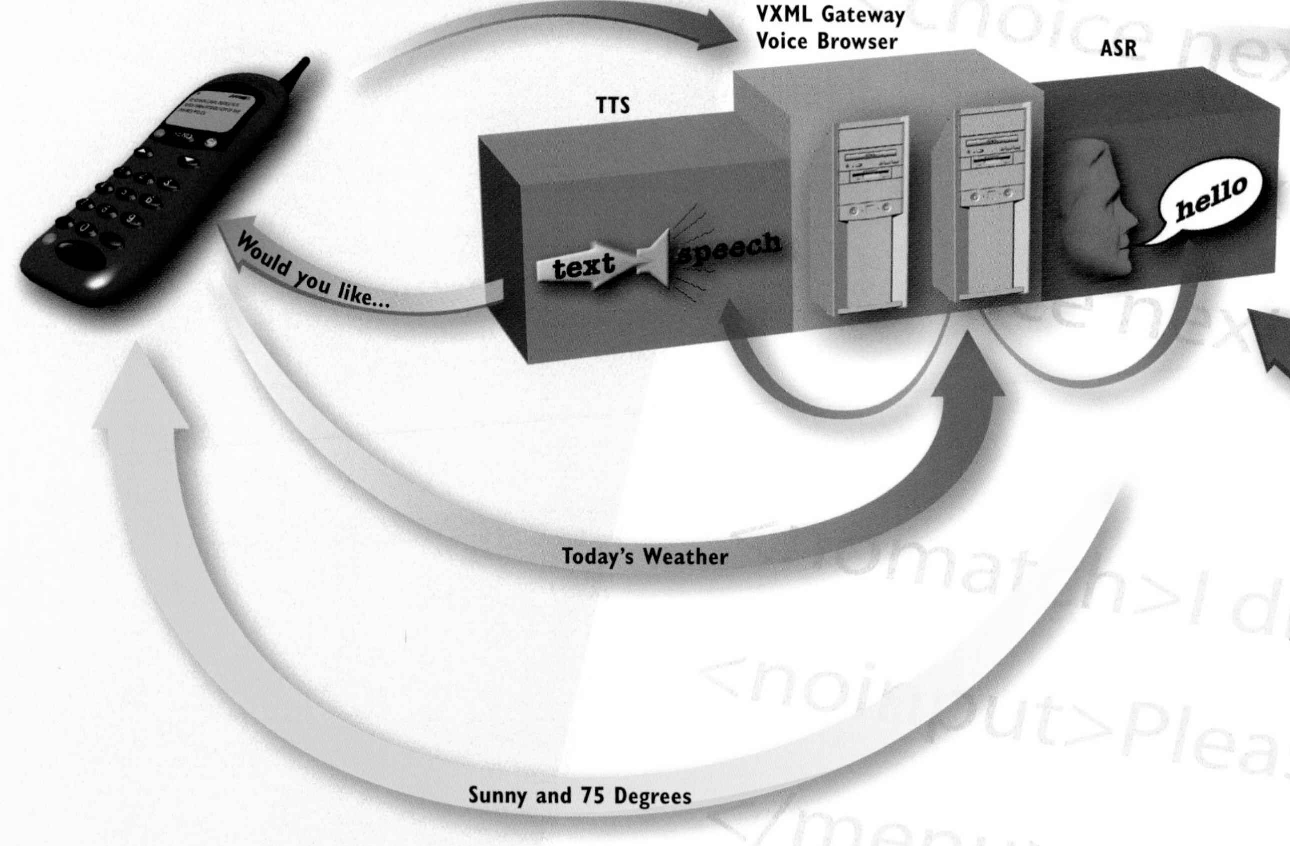

4 Based on the number the phone dialed, the VXML gateway requests a VXML document from a Web server.

5 The server delivers the VXML document to the gateway, which interprets the document and instructs the TTS component to read the document or choices on the document menu to the caller.

6 The caller listens to the menu and speaks into the phone, or uses the phone keypad to make his choice.

7 The VXML gateway uses the ASR component to interpret the caller's requests and, using the voice browser, inputs that information into the VXML document. The document delivers, via the voice browser in the gateway, the information the caller has requested.

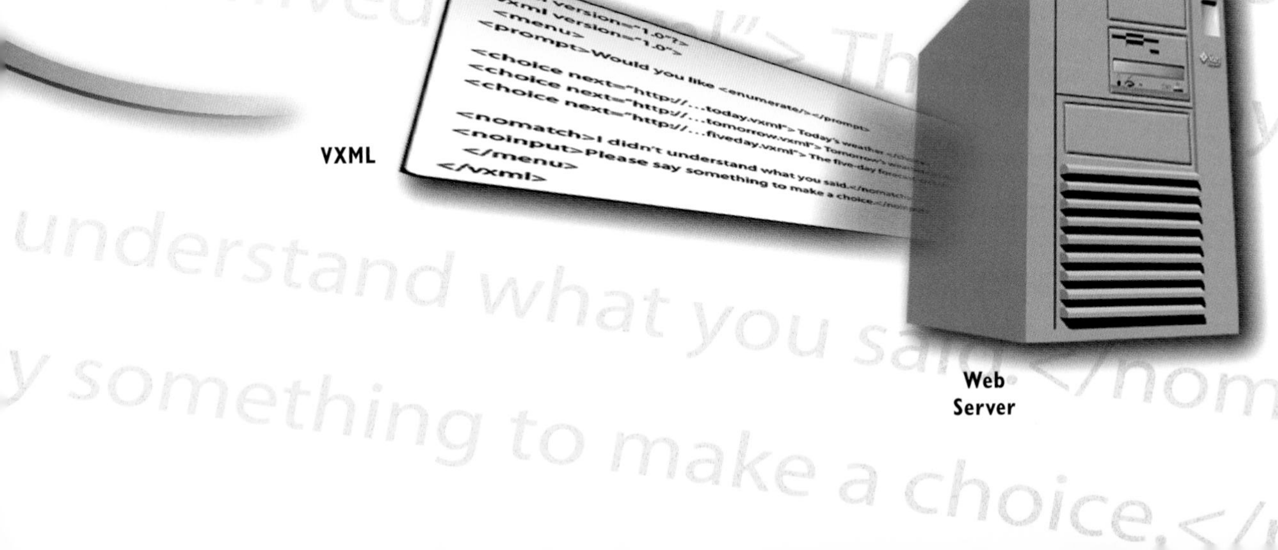

CHAPTER 19 How i-mode Works

THE United States hasn't been in the forefront of accessing the Web, e-mail, and other Internet and interactive services with cell phones. Although there has been a lot of talk about it, it's still not a daily part of many people's lives. There are several reasons for that, including relatively high costs, a relatively small installed base of phones that can access the Internet, low connection speeds, and, to a great extent, no compelling reason for people to hop onto the Internet with their telephones.

That's not the case, however, in the rest of the world. In Japan, cell phone access to the Internet has become something of a national craze, with teenagers as well as businesspeople regularly getting a variety of services and information through their cell phones.

In the United States, a primary way that people access the Internet is through the Wireless Access Protocol (WAP). For more information about how WAP works, turn to Chapter 17, "How Cell Phones Access the Internet."

In Japan, though, people access the Internet a different way, using a service called i-mode. i-mode services offer much more content and services than do cell phones in the United States—everything from e-mail and chat to stock quotes, online shopping, games, quizzes, and horoscopes. To give you just a small example of how popular the service is, as of months ago, some 600,000 people used a service that offers daily quizzes and horoscopes, provided by a company called Bandai. And the Sanwa Bank in Japan estimates that more people bank using mobile phones than bank using PCs.

In Japan, a single company runs i-mode—it's not like in the United States where many different companies compete to provide WAP cell phone service. The Japanese firm NTT DoCoMo is in charge of the i-mode service, and it has plans to bring that service to the rest of the world. Its first target is Europe, and after that, it might bring it to the United States.

Unlike WAP, i-mode is an always-on service. It delivers information at a relatively low speed—9,600 bits per second (bps). But it won't stay at that slow speed for long. DoCoMo has been building an infrastructure to use 3G cellular technology (the new generation of high-speed Internet cellular access) with i-mode. There have been delays in getting it off the ground, which is no great surprise, because there have been delays worldwide in delivered 3G cellular services. (For more information about how 3G works, turn to Chapter 24, "How Wireless 3G Works.") But if it's not rolled out by the time you read this, it will be available soon.

How i-mode Works

1. To use the i-mode service, you must use a special i-mode phone—normal cell phones won't work. i-mode phones are popular in Japan, and as of this writing aren't available in the United States. The phones include a special Web microbrowser designed specifically for the i-mode service. Their screens also are slightly larger than other cell phones, ranging from 96×108 pixels to 120×130 pixels, which means they can display from 6–10 lines of text and 16–30 characters of text per line. The screens can be monochrome or grayscale, or they can display up to 256 colors, and most can show small animations in the animated GIF format.

2. Unlike other cell phones, i-mode phones are always connected and online—you don't need to dial them to make a connection to get or send information. They use General Packet Radio Service (GPRS) technology to keep this always-on connection. So, for example, when you've received e-mail, you'll be notified instantly; you won't have to dial in to check. This also means that the phones can deliver services such as instant messaging. Because of this always-on connection, i-mode subscribers are charged according to the amount of data transmitted, not by connection time.

6. The i-mode gateway sends the page, e-mail, or other information to the i-mode cell phone. The connection speed is 9,600 bps—a relatively slow speed when compared to computer connections. However, i-mode sites are very small, and average only about 1.2 kilobytes in size, so the download usually takes only a few seconds. E-mails are limited to 500 bytes; if an e-mail is larger than that, only the first 500 bytes will be transmitted and the rest can't be read.

7. The microbrowser in the i-mode phone displays the information, Web site, or e-mail. Because of the large numbers of subscribers and the great demand for i-mode, the kinds of sites, information, and services available through i-mode are astonishing—everything from games and animations to instant messaging, banking and stock information, weather, astrology, recipes, reference tools, and far more.

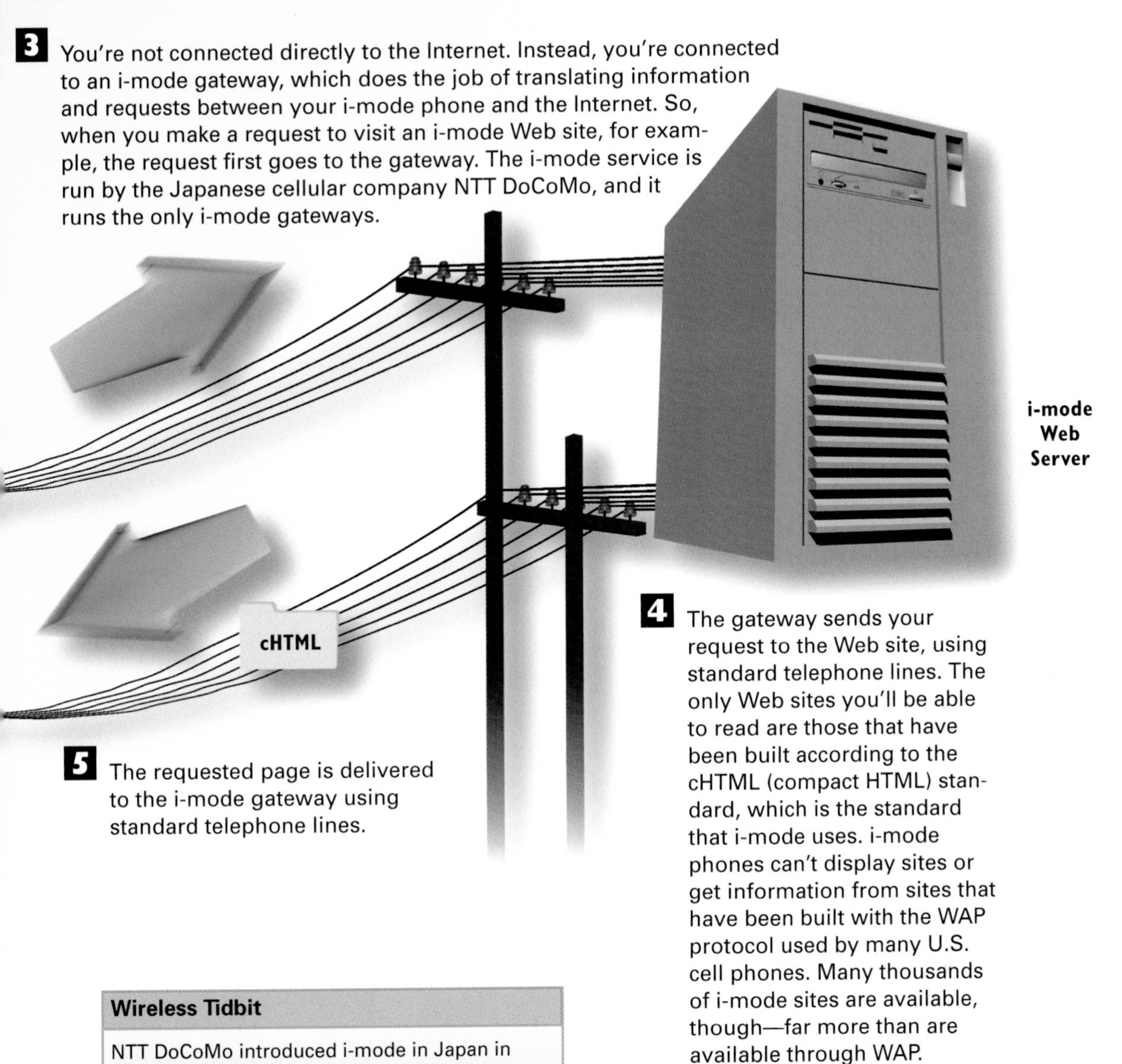

3 You're not connected directly to the Internet. Instead, you're connected to an i-mode gateway, which does the job of translating information and requests between your i-mode phone and the Internet. So, when you make a request to visit an i-mode Web site, for example, the request first goes to the gateway. The i-mode service is run by the Japanese cellular company NTT DoCoMo, and it runs the only i-mode gateways.

4 The gateway sends your request to the Web site, using standard telephone lines. The only Web sites you'll be able to read are those that have been built according to the cHTML (compact HTML) standard, which is the standard that i-mode uses. i-mode phones can't display sites or get information from sites that have been built with the WAP protocol used by many U.S. cell phones. Many thousands of i-mode sites are available, though—far more than are available through WAP. i-mode gateways also channel requests for e-mail and for visiting Web sites.

5 The requested page is delivered to the i-mode gateway using standard telephone lines.

Wireless Tidbit

NTT DoCoMo introduced i-mode in Japan in February 1999, and its growth since then has been astonishing. In only a little more than two years, by April 2001, the number of i-mode subscribers had nearly reached 23 million.

CHAPTER 20 PCs and Wireless Technology

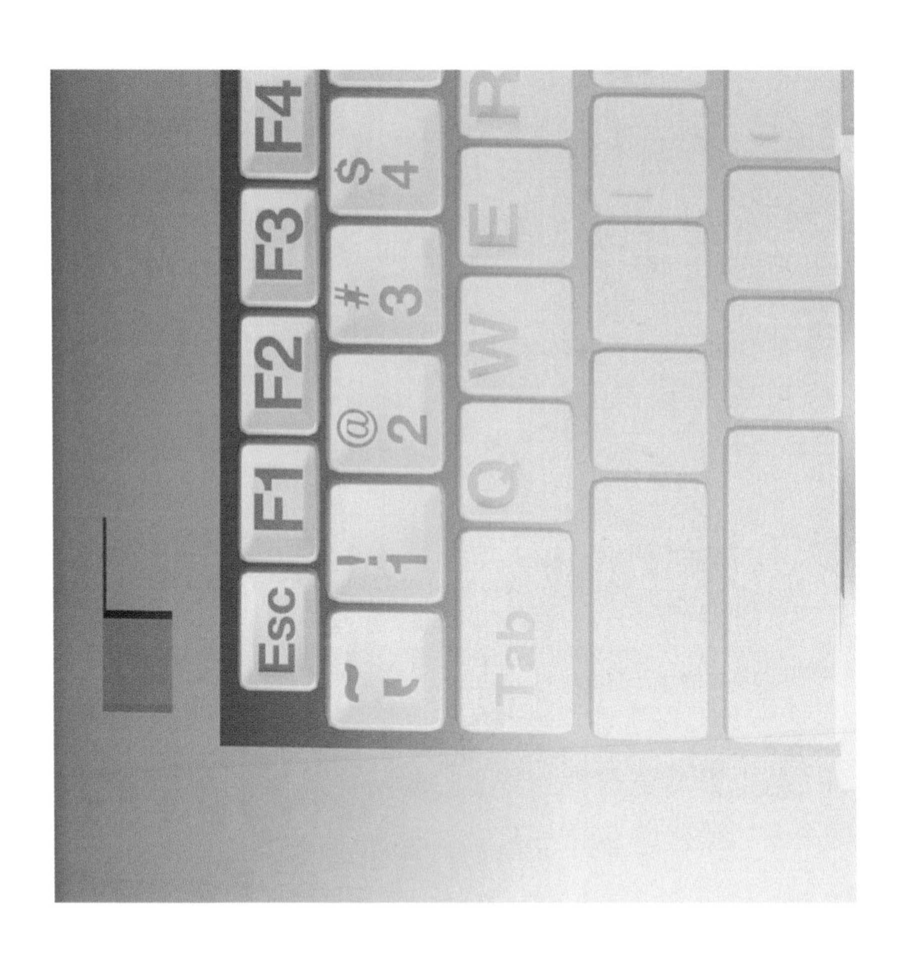

FOR most of this book, you've seen how wireless technologies are used for various kinds of communications—between cellular phones, over the Internet, and among computer networks, among many others. But wireless technologies can be used not just to communicate between different devices. They can also be used to help devices themselves work. The most obvious example is the remote control, which helps you use your television. Many other examples abound, however, from garage door openers to remote control toys. Wireless technology can be used to help computers operate as well. For now, these technologies are not that commonly used, although they are slowly catching on.

If you own a laptop computer, you might notice a curious, small, dark, red piece of plastic on it somewhere. (It is on some desktop computers, as well.) It's usually quite unobtrusive and blends in with its surroundings. That plastic protects the *IrDA* (*Infrared Data Association*) port, a port that can be used to communicate with other computers or with devices such as printers.

It's safe to say that this port is rarely used. In fact, many computers come with the port disabled; to use it, you'll have to navigate through a series of menus and check off the proper boxes on the right screen. It communicates with computers and devices using a technology similar to your TV's remote control. Both use infrared, rather than RF waves, to communicate. Your computer and the computer or device with which you're communicating need to be within one meter of each other, and they need to be in a direct line with each other with no obstacles in the way. The most common use of this technology is to print to infrared-enabled printers, although in theory it also can be used to communicate with other PCs.

A more popular way to use wireless technologies in PCs is for wireless keyboards and mice. Unlike PC infrared technology, which has a standard to adhere to (IrDA), there is no standard way for wireless keyboards and mice to work. They can use infrared or RF frequencies.

The best wireless devices use RF rather than infrared. With RF, there's no need for the mouse or keyboard to point directly at an infrared port; instead, a radio transmitter in the device sends out signals that are picked up by the PC. They only need to be within six feet of one another. Given all the obstacles on a typical computer user's desk, it's a good thing that a clear line of sight is not needed, because it's rare that there will ever be one.

How Wireless Mice and Keyboards Work

6 The computer acts on the signal—for example, it displays the letter A on the monitor.

Wireless Tidbit

With Logitech wireless keyboards and mice, the maximum distance at which the receiver and transmitter can communicate with one another is approximately six feet, so wireless mice and keyboards farther away than six feet won't interfere with it. However, if any mice or keyboards are within six feet, they won't interfere, either. A 12-digit ID is assigned to every keyboard and mouse and its accompanying receiver. The receiver will only accept signals from devices with that ID, and it ignores all other signals from other wireless keyboards and mice.

3 When you press a key on the keyboard or move the mouse, the device creates a digital signal as it normally does—it translates the letter A, for example, into a keyboard code that the computer understands as the letter A. In this case, the keyboard code for A is 1E.

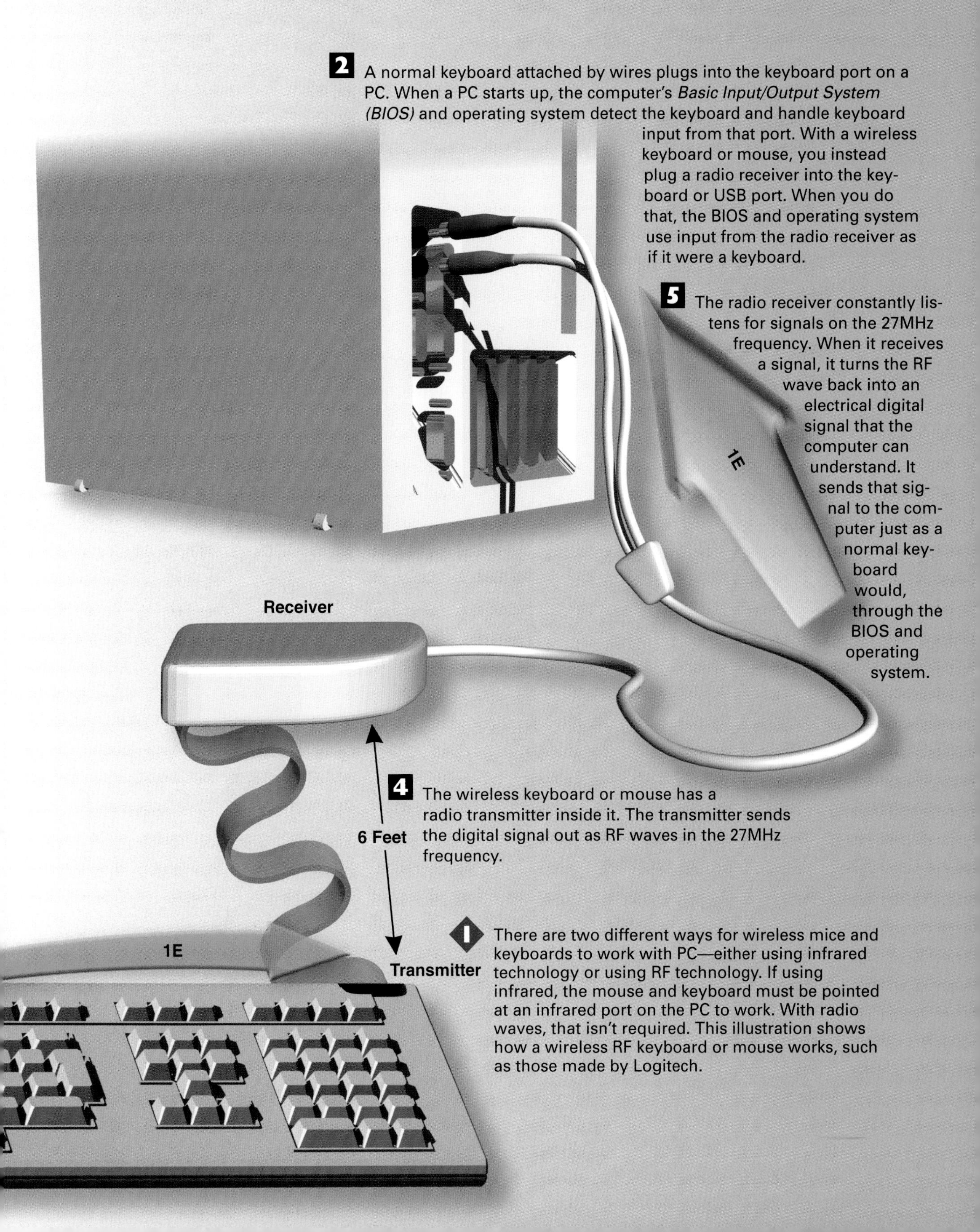

2 A normal keyboard attached by wires plugs into the keyboard port on a PC. When a PC starts up, the computer's *Basic Input/Output System (BIOS)* and operating system detect the keyboard and handle keyboard input from that port. With a wireless keyboard or mouse, you instead plug a radio receiver into the keyboard or USB port. When you do that, the BIOS and operating system use input from the radio receiver as if it were a keyboard.

5 The radio receiver constantly listens for signals on the 27MHz frequency. When it receives a signal, it turns the RF wave back into an electrical digital signal that the computer can understand. It sends that signal to the computer just as a normal keyboard would, through the BIOS and operating system.

4 The wireless keyboard or mouse has a radio transmitter inside it. The transmitter sends the digital signal out as RF waves in the 27MHz frequency.

1 There are two different ways for wireless mice and keyboards to work with PC—either using infrared technology or using RF technology. If using infrared, the mouse and keyboard must be pointed at an infrared port on the PC to work. With radio waves, that isn't required. This illustration shows how a wireless RF keyboard or mouse works, such as those made by Logitech.

How Infrared Printing Works

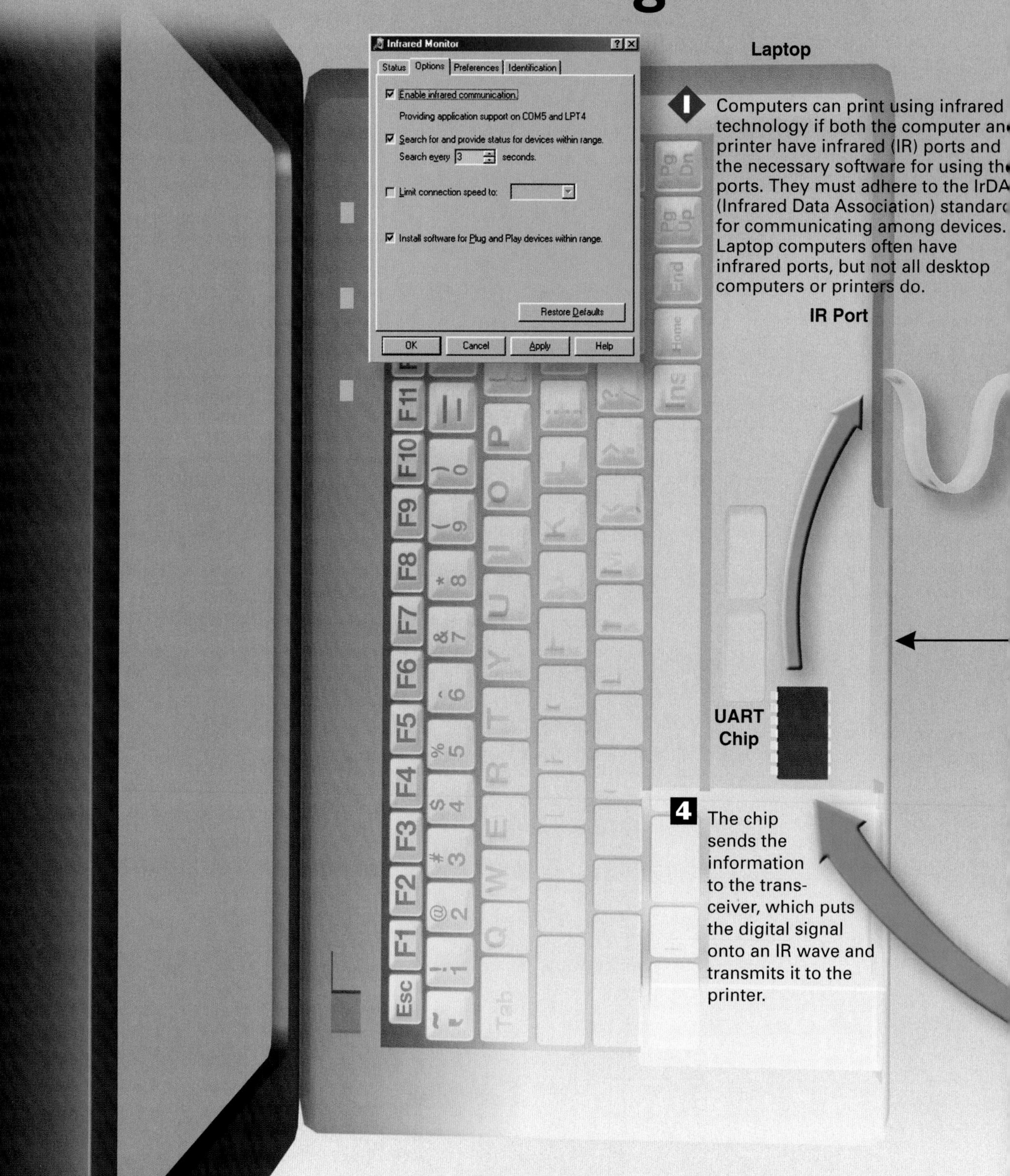

1 Computers can print using infrared technology if both the computer and printer have infrared (IR) ports and the necessary software for using the ports. They must adhere to the IrDA (Infrared Data Association) standard for communicating among devices. Laptop computers often have infrared ports, but not all desktop computers or printers do.

4 The chip sends the information to the transceiver, which puts the digital signal onto an IR wave and transmits it to the printer.

6 The printer's transceiver receives the printer data through the IR wave and turns it into a digital electric signal that the printer can understand. It sends that signal to the printer, just as if it were a printer connected to a computer by a cable, and prints the page.

5 The PC and printer communicate over the IR wave using what is called the *IrDA protocol stack*. This is a set of protocols—rules—that details how IR devices talk to one another.

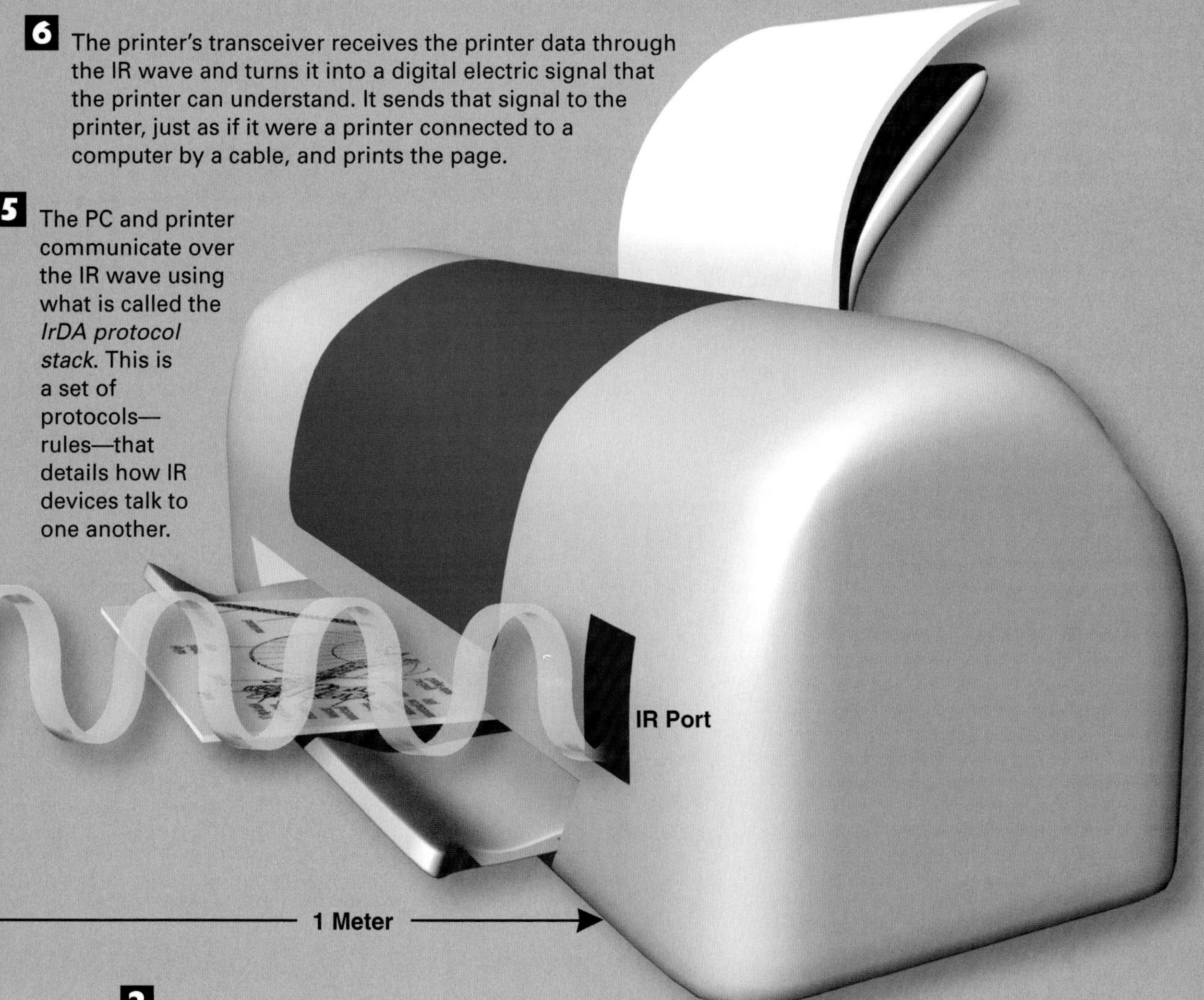

2 Infrared requires that there be a clear line of sight between infrared devices—the ports must be in line with each other, with nothing blocking them. That's because infrared can't travel through or around obstacles. They also must be within a meter of each other. Both the printer and computer have transceivers in them, devices that can send and receive using infrared rays.

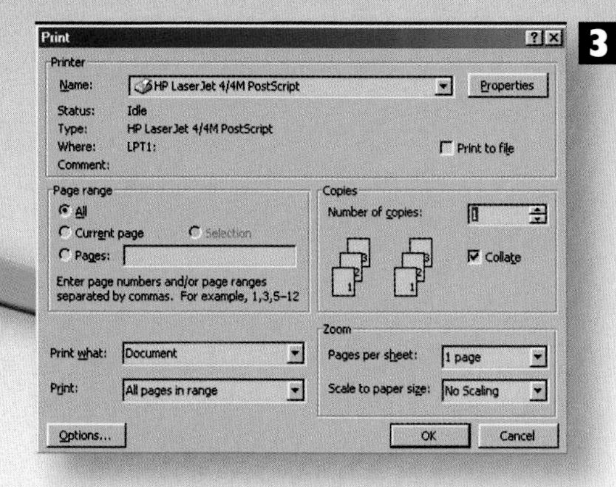

3 When someone wants to print, he issues a print command, no differently from any other print command. The command goes through a printer driver, which handles the work of sending information to the printer. It sends the information to a *UART* (Universal Asynchronous Receiver and Transmitter) chip.

CHAPTER 21 How Wireless Palmtops Work

ALTHOUGH wireless access to the Internet is only beginning to catch on for cell phone users, there's a small, portable device that has been used for years for wireless access to the Internet: palmtops, often referred to as Personal Digital Assistants (PDAs). These palmtops are used primarily as a way for people to keep track of personal and business information. They include an address book, a calendar, a memo page, a To-Do list, and similar productivity-based programs. People also can download and install thousands of other programs onto them, from games to databases and beyond.

Generally, there are two kinds of these devices—those made by Palm and that use the Palm operating system, and those that use Microsoft's Windows CE operating system. Palm devices are by far the more popular of the two.

The first palmtops didn't offer wireless access to the Internet, but the second generation started to offer wireless access, either by being built directly into the device, or by means of a special wireless modem that attaches to the device. Palm makes a model, called the Palm VII, that includes an antenna and built-in radio for wireless Internet access. It uses a cellular network for Internet access that works very much like the cellular network for cell phones.

There are two problems with giving a device like the Palm VII access to the Internet. One is that the cellular network that it uses is a low-speed network, so receiving Web pages over it can take a substantial amount of time. The second problem is that the small screen of the Palm isn't well-suited for displaying Web pages—the screen is far too small, doesn't offer a high resolution, and many Palms don't use color.

To solve both those problems, Palm devised a technique called *Web clipping*. With Web clipping, you don't actually get Web pages delivered to you. Instead, you get information sent to you from the Web pages, and the information is formatted especially for the Palm's screen. Software called a *Palm Query Application (PQA)* runs on the Palm, and data from the Web is delivered into that software in a form easy for you to read.

Palms and other palmtops also let you send and receive e-mail. To solve the problem of low-speed access, you get only a screenful of information at a time—if you want more, you ask for a second screen, and so on. Attachments are stripped out because the Palm can't handle them.

Palms and other palmtops also can use add-on software to browse the Web using the WAP protocol, which was designed for cell phones. For more information about WAP, turn to the illustration "How Web Pages Are Delivered with WAP," in Chapter 17, "How Cell Phones Access the Internet."

There is one more way in which some palmtops—notably, the Palm—use wireless technology. They allow palmtop users to beam data and programs directly to each other using a built-in infrared port.

A Cutaway View of a Wireless Palmtop

Infrared port Most palmtops have an infrared port that's used to exchange information and programs with other palmtops, and in some instances, with computers as well.

RAM (Random Access Memory) These memory chips store all your data, such as your memos, addresses, and calendar information. They also store any extra programs you install into your palmtop beyond what came preinstalled. Palmtops usually have at least 2MB of RAM, and some come with 32MB or more. Unlike with PCs, when you turn off your palmtop, data and programs aren't lost from RAM. That's because even when your palmtop is shut off, the device still draws a tiny bit of power from the batteries, and feeds that power to RAM so that data and programs aren't lost.

ROM (Read-Only Memory) These memory chips hold the palmtop's operating system, as well as its accompanying preinstalled software, such as a calendar, address book, and memo pad. The operating system and software stay in ROM even when the palmtop's power is turned off—it's there permanently.

Microprocessor This is the brains of a palmtop. It does all the processing, shuttling information into and out of memory, managing wireless and other communications, accepting data input, and all other similar chores. It's a much smaller, less-expensive, and less-powerful microprocessor than you'll find on desktop or laptop computers.

USB or serial port Palmtops are designed to sync their data with data on your PC. That way, both your PC and your palmtop have the same address information, memos, calendar, and similar information. The palmtop usually attaches to the PC to sync either through a USB or serial port.

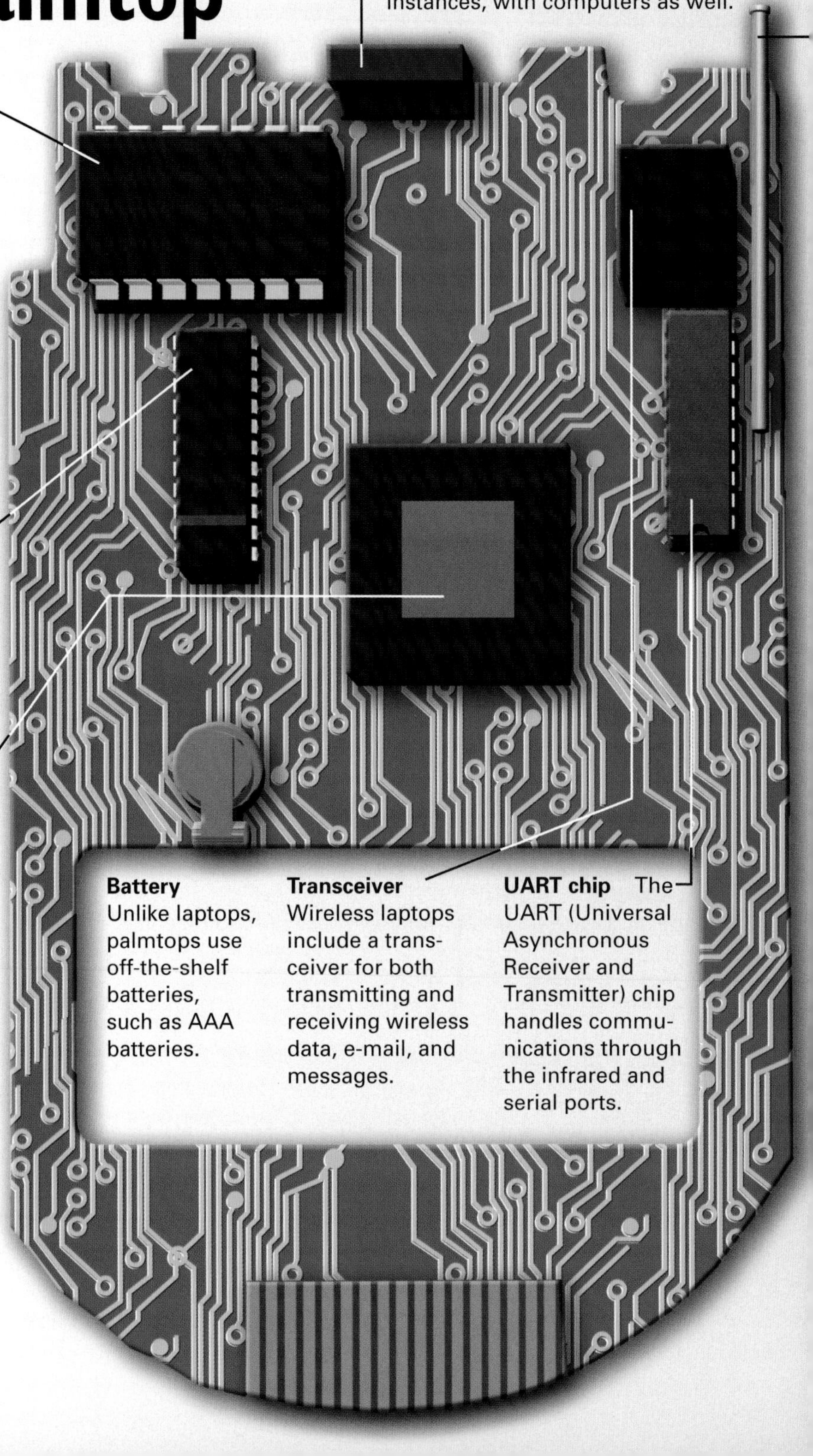

Battery Unlike laptops, palmtops use off-the-shelf batteries, such as AAA batteries.

Transceiver Wireless laptops include a transceiver for both transmitting and receiving wireless data, e-mail, and messages.

UART chip The UART (Universal Asynchronous Receiver and Transmitter) chip handles communications through the infrared and serial ports.

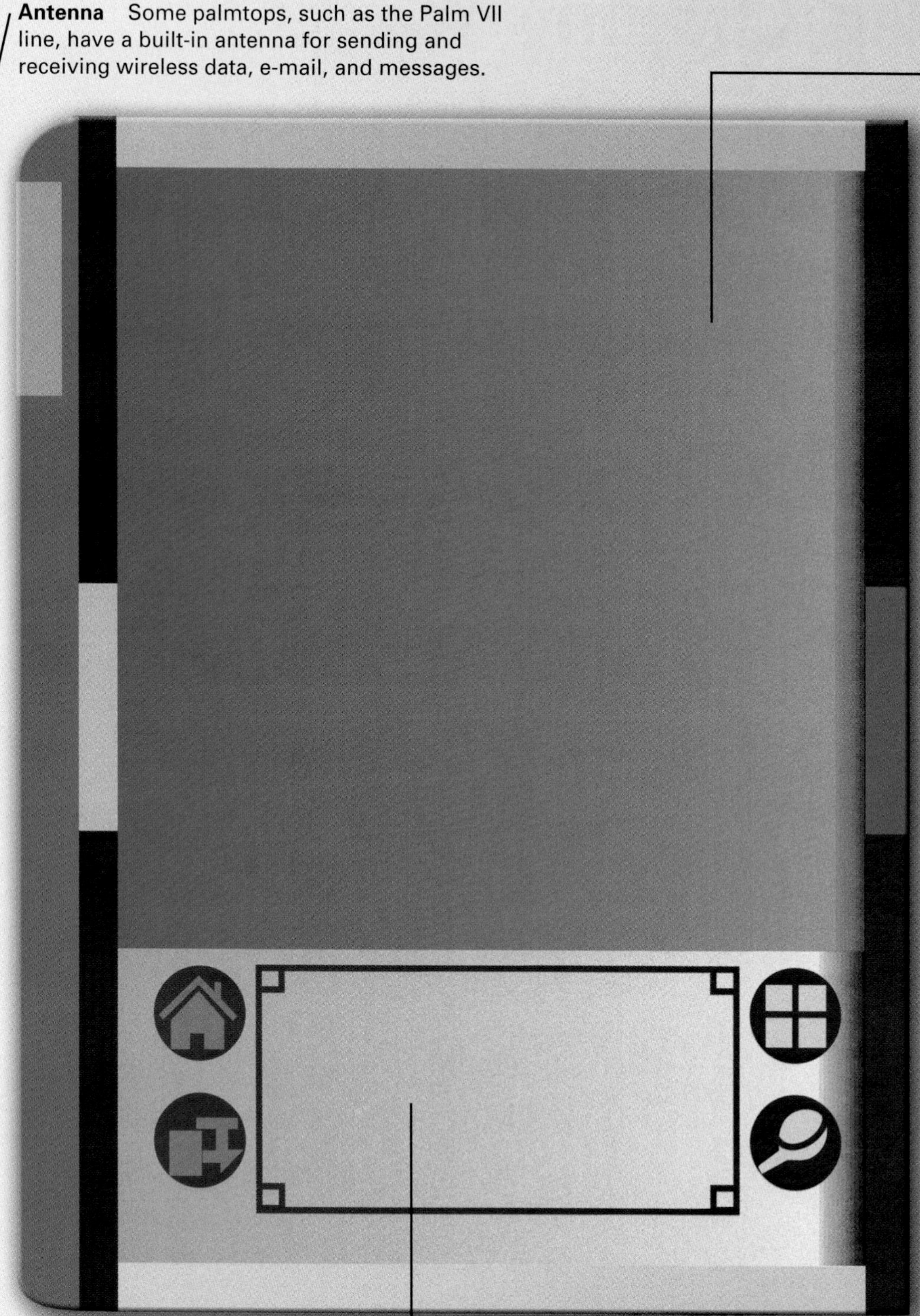

Antenna Some palmtops, such as the Palm VII line, have a built-in antenna for sending and receiving wireless data, e-mail, and messages.
LCD display This is the equivalent of a computer's monitor. It's where you see all your data and programs. The quality of palmtop LCDs varies greatly; some have low-resolution black-and-white and others have high-resolution color.
Handwriting recognition area Many palmtops, notably the popular Palm line and Palm-compatibles, use handwriting recognition as a way to accept input. You write on this small area, and the palmtop recognizes what you've written and uses that as input.

Wireless Tidbit

Although many different kinds of palmtops are available, the most popular are those from Palm and that use the Palm operating system. More than 70 percent of all palmtops are Palm or Palm-compatible.

How Web Clipping Works

1 *Web clipping* is the term used to describe the way in which the wireless Palm VII and similar devices get information from the Internet. Other Palm devices also can use wireless modems to get information from the Internet through Web clipping.

2 To get information from the Internet, a piece of software called a *Palm Query Application (PQA)* must run on the Palm. Each site that delivers information to the Palm VII has its own PQA; you can't use the PQA for Yahoo!, for example, to get information from eTrade.

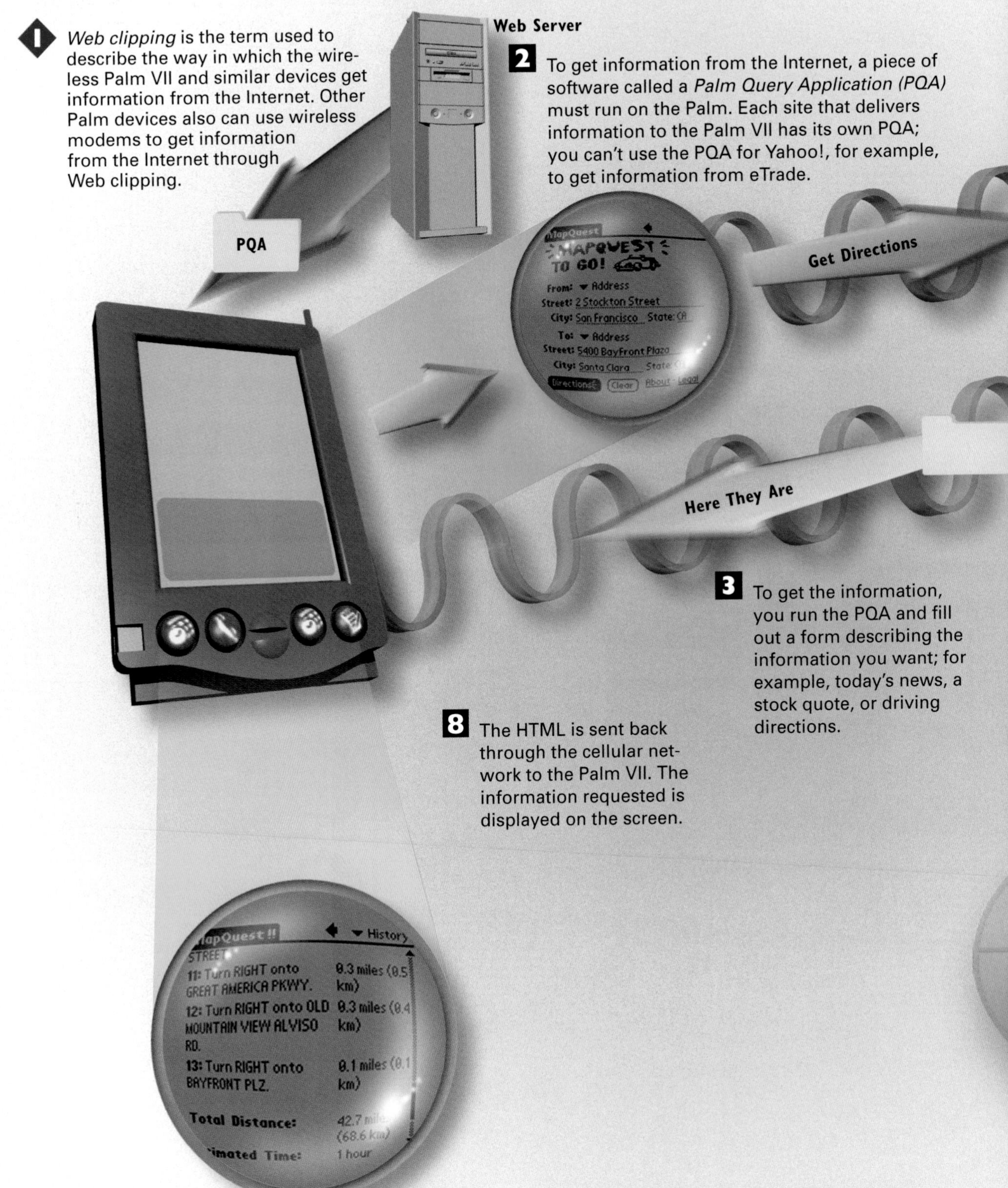

3 To get the information, you run the PQA and fill out a form describing the information you want; for example, today's news, a stock quote, or driving directions.

8 The HTML is sent back through the cellular network to the Palm VII. The information requested is displayed on the screen.

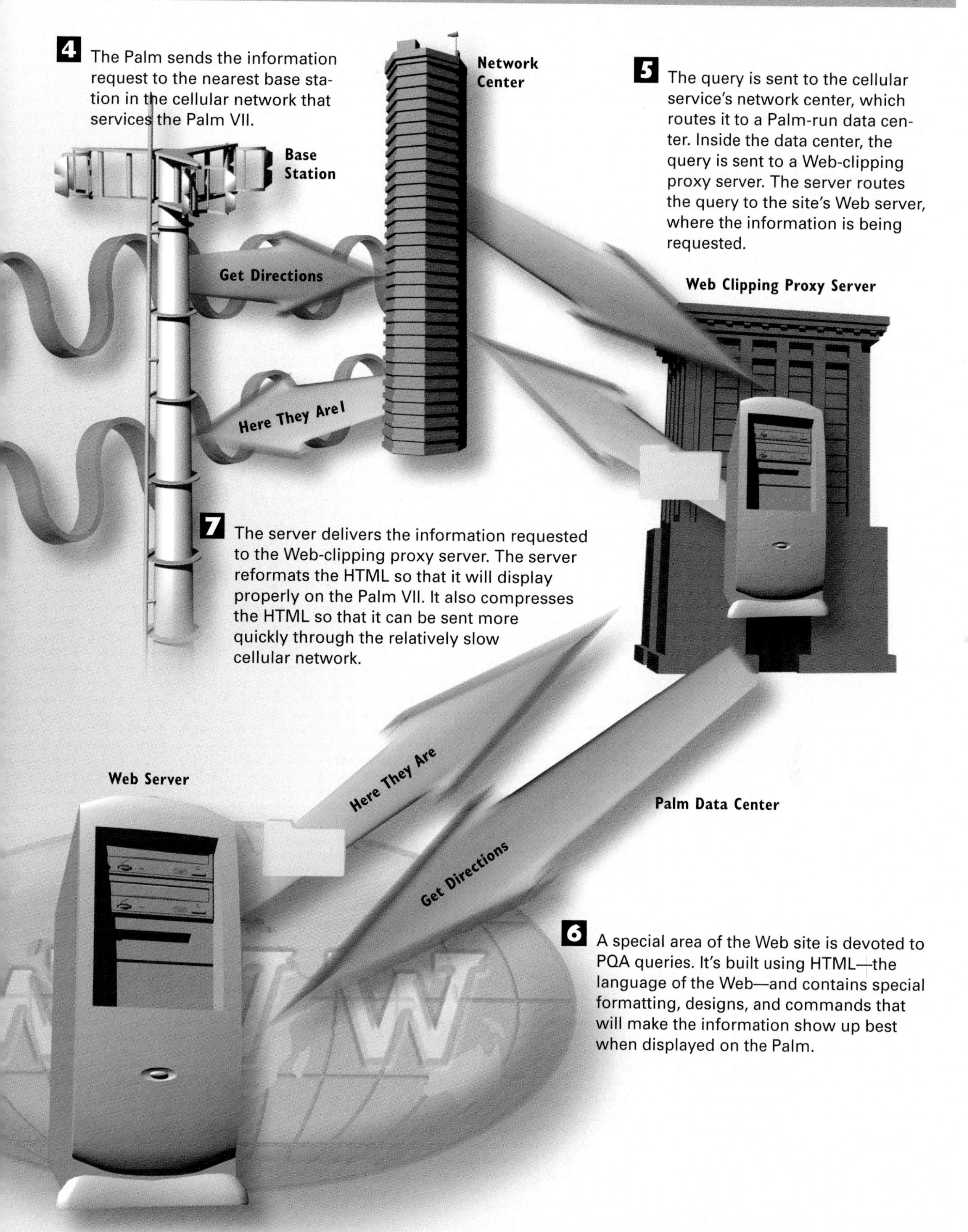

4 The Palm sends the information request to the nearest base station in the cellular network that services the Palm VII.

5 The query is sent to the cellular service's network center, which routes it to a Palm-run data center. Inside the data center, the query is sent to a Web-clipping proxy server. The server routes the query to the site's Web server, where the information is being requested.

6 A special area of the Web site is devoted to PQA queries. It's built using HTML—the language of the Web—and contains special formatting, designs, and commands that will make the information show up best when displayed on the Palm.

7 The server delivers the information requested to the Web-clipping proxy server. The server reformats the HTML so that it will display properly on the Palm VII. It also compresses the HTML so that it can be sent more quickly through the relatively slow cellular network.

How Wireless Palmtops Send and Receive E-Mail

How iMessenger E-Mail Works

1 Depending on the wireless palmtop you have, it will send and receive e-mail differently. This illustration shows the two primary ways that a wireless Palm VII can send and receive e-mail. The first way is to use the built-in iMessenger service. To receive mail, you first run the iMessenger software.

2 You'll connect to the cellular network through the base station, and then to the network operations center.

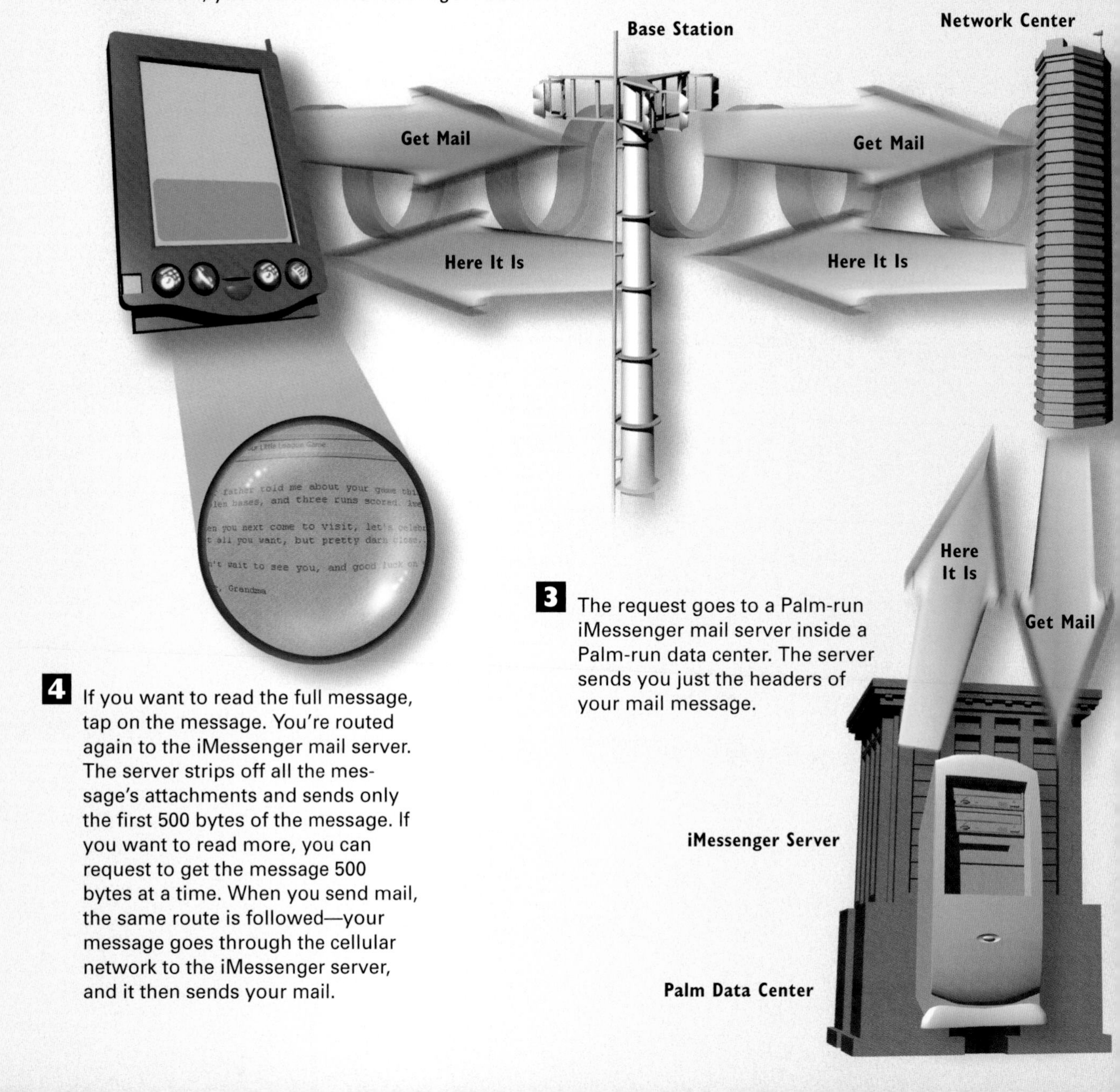

3 The request goes to a Palm-run iMessenger mail server inside a Palm-run data center. The server sends you just the headers of your mail message.

4 If you want to read the full message, tap on the message. You're routed again to the iMessenger mail server. The server strips off all the message's attachments and sends only the first 500 bytes of the message. If you want to read more, you can request to get the message 500 bytes at a time. When you send mail, the same route is followed—your message goes through the cellular network to the iMessenger server, and it then sends your mail.

How Palms Retrieve Mail from Internet Mail Accounts

1 You also can use the Palm VII to retrieve mail from your normal Internet mail account. To do that, you first download and install a piece of software called ThinAir Mail.

2 Next, you configure it to work with your Internet mail account, by putting in information such as your POP3 mail address, username, and password. When you want to receive mail, you run the software.

3 It connects to the cellular network through the base station, then to the network operations center, and then to the Internet, to a proxy mail server run by ThinAir Mail.

4 The proxy server sends your request to get mail to your Internet mailbox on a POP3 server. Your mailbox sends the headers of your messages to the ThinAir Mail proxy, which reformats the information so it can be easily read on the Palm screen. It then sends that information back to you. You choose which mail to read. The request goes back through the cellular network and to the ThinAir Mail proxy server, and from there to your mail server. You're sent the first 500 bytes of the message, and can retrieve more 500 bytes at a time if you want.

5 You send mail in a similar manner—connecting through the cellular network to the ThinAir Mail proxy server. The server then sends your mail to your *Simple Mail Transfer Protocol (SMTP)* server, which sends your mail.

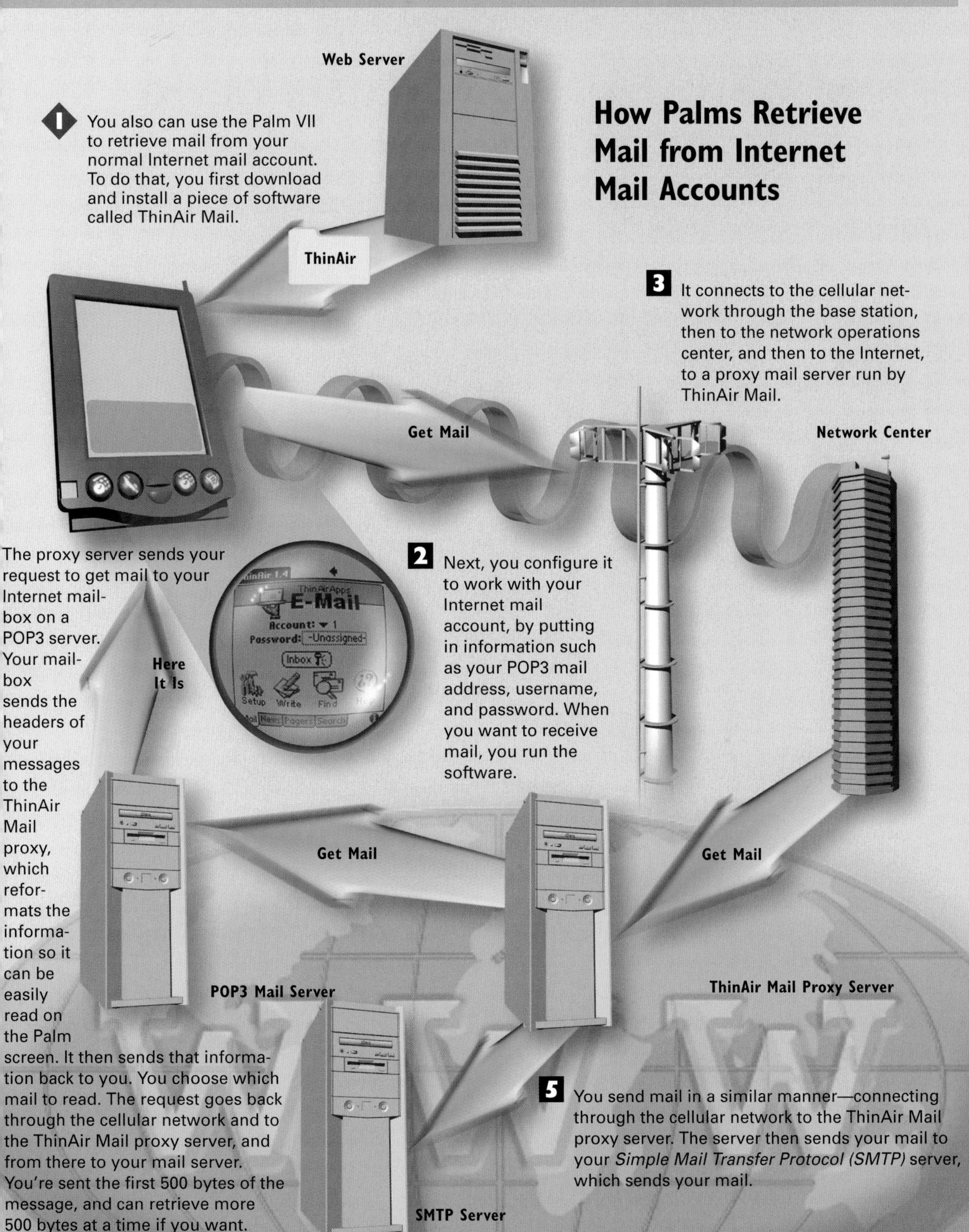

How Palmtops Beam Data to Each Other

4 When you want to beam data—for example, your address, phone number, e-mail address and other personal information, or a document of some kind—you run the program that has the data in it, and then choose the Beam function from a menu.

1 The most popular palmtop is the Palm, so this illustration shows how Palms beam data and applications to one another. Palms have an infrared port at the top of the device that is used to exchange data and programs with other Palms. It includes a transceiver for transmitting and receiving data and programs using infrared rays.

5 The UART shuttles the data from the application to the infrared port and tells it to beam the data.

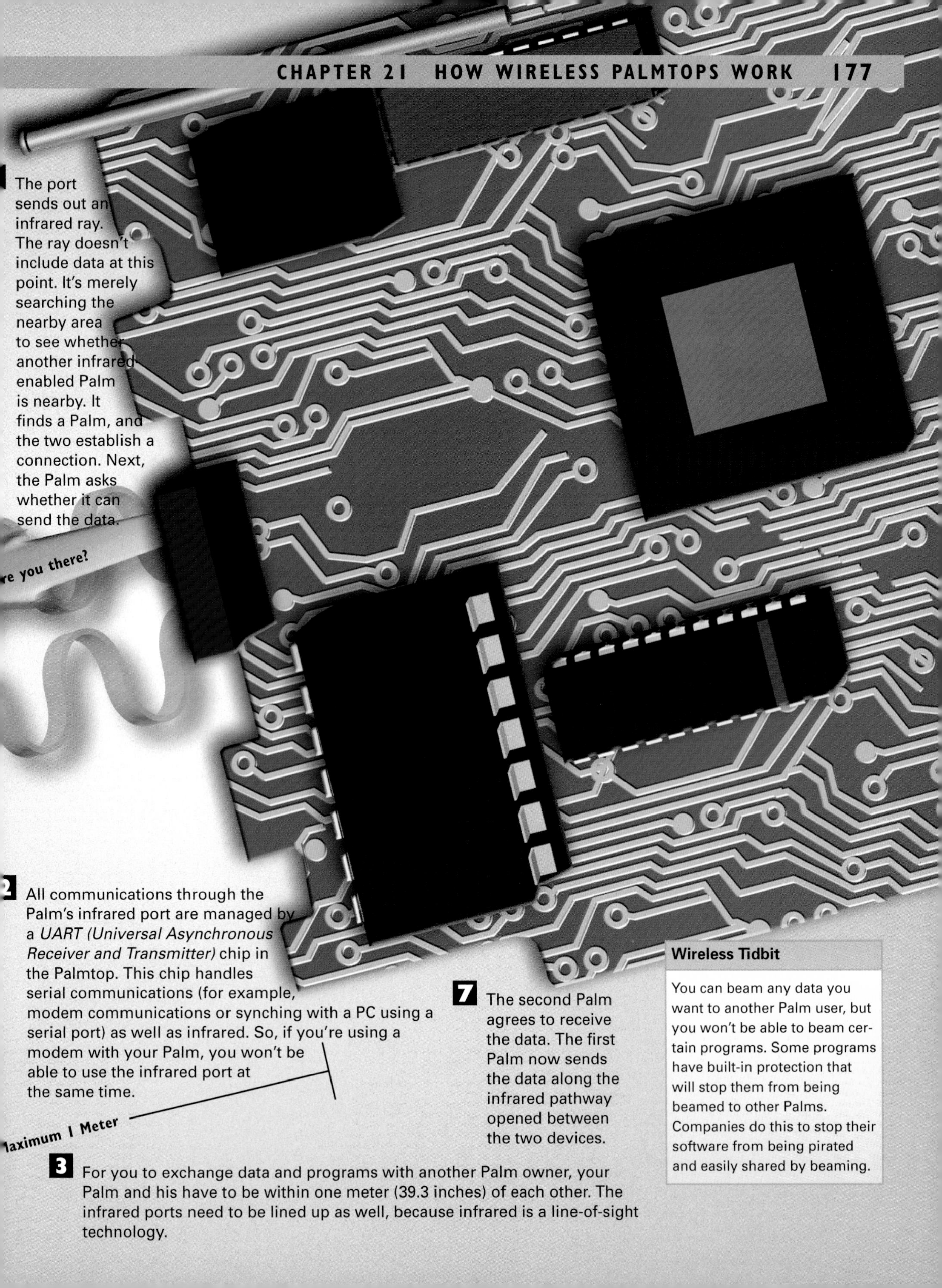

1 The port sends out an infrared ray. The ray doesn't include data at this point. It's merely searching the nearby area to see whether another infrared-enabled Palm is nearby. It finds a Palm, and the two establish a connection. Next, the Palm asks whether it can send the data.

2 All communications through the Palm's infrared port are managed by a *UART (Universal Asynchronous Receiver and Transmitter)* chip in the Palmtop. This chip handles serial communications (for example, modem communications or synching with a PC using a serial port) as well as infrared. So, if you're using a modem with your Palm, you won't be able to use the infrared port at the same time.

3 For you to exchange data and programs with another Palm owner, your Palm and his have to be within one meter (39.3 inches) of each other. The infrared ports need to be lined up as well, because infrared is a line-of-sight technology.

7 The second Palm agrees to receive the data. The first Palm now sends the data along the infrared pathway opened between the two devices.

Wireless Tidbit

You can beam any data you want to another Palm user, but you won't be able to beam certain programs. Some programs have built-in protection that will stop them from being beamed to other Palms. Companies do this to stop their software from being pirated and easily shared by beaming.

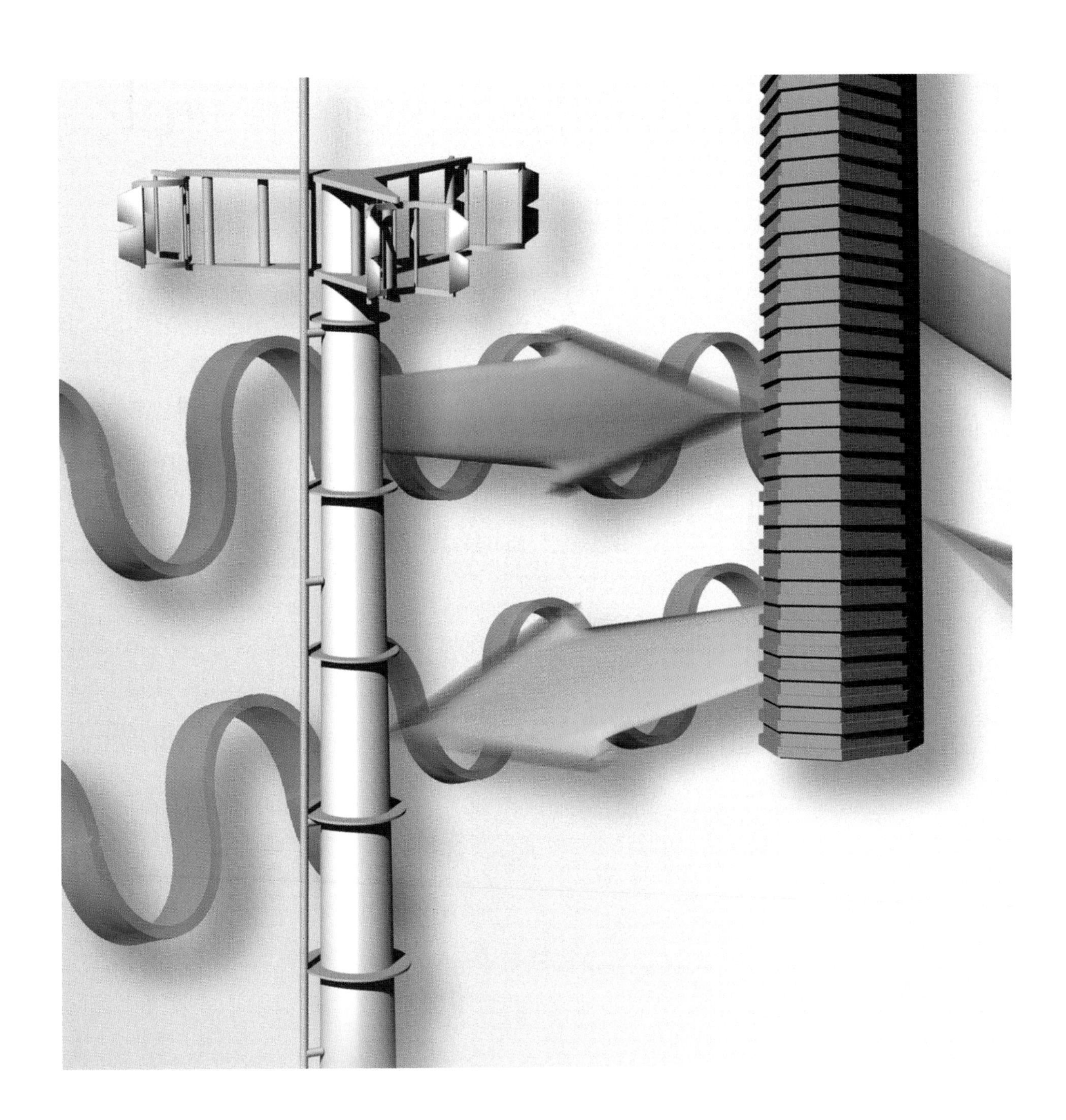

PART

Applying Wireless Technology: mCommerce, Security, Business Use, and Beyond

CELL phones or other wireless technologies can be entertaining, but they also must have some real uses. The whole point of using wireless technology, after all, is to reach people, or do business, or in some way to make your life easier or more convenient. In this part of the book, we'll look at a variety of ways in which wireless technology is used. And we'll look at a variety of miscellaneous uses of the technology, as well as what the future of wireless technology will look like.

Chapter 22, "How mCommerce and Corporate Wireless Access Work," looks at the most important ways that wireless technology is used in corporations and for buying and selling. It starts off by examining mCommerce, the term commonly used for mobile commerce. As of yet, mCommerce is not yet big, but expectations are that it will boom in the coming years. According to one estimate, in fact, several billion dollars a year will be spent buying and selling over cell phones. The chapter looks at a wireless commerce standard called Mobile Electronic Transactions (MET) that uses encryption to be sure you can buy and sell without someone stealing your credit card. And it shows you how a variety of technologies, such as the Wireless Access Protocol (WAP), the Wireless Transport Layer Security, WMLScript, and others will make it easy and safe for you to buy from your cell phone.

The chapter also looks at how corporations can use wireless technologies to give employees access to all of a company's resources, no matter where they are. A salesperson, for example, could get up-to-the-minute pricing information and availability for goods or services, and could place an order directly on a personal digital assistant. The systems to allow this apparently simple matter are quite complex, as you'll see in this chapter.

Chapter 23, "Privacy and Security in a Wireless World," looks at cellular's seamy underbelly and at privacy issues anyone who uses wireless communication should know about. You'll learn all about a variety of dangers in this chapter: For example, you might not realize it, but it's exceedingly simple for someone to listen to your cellular calls with inexpensive, off-the-shelf hardware. The chapter takes a close look at how snoopers can listen in to your cell phone calls.

Snoopers can do more than that—with a laptop computer, the right hardware, and a little know-how, they can tap into all the traffic going through a wireless network, and can do it from the parking lot outside a company. And there are many more dangers, as well. Viruses have already targeted PDAs, and the first forms of a cell phone virus have been circulating as well. In this chapter, you'll find a detailed explanation of how these cellular viruses work.

The chapter looks at another danger you might not know about—the "cloning" of cell phones. Some cell phones can be cloned, and then sold, allowing other people to make calls using your cell phone, and leaving you to foot the bill.

Chapter 24, "How Wireless 3G Works," looks at one of the most talked-about technologies in the wireless world—3G (for third-generation) technology. This technology will allow very high-speed access to the Internet, allowing for things such as streaming videos and music straight to your cell phone. The technology has been slower in coming than many people had hoped, due largely to financial problems that wireless providers have faced, as well as some technical hurdles. But no one doubts that 3G will be here soon, and will eventually become the dominant way that people use cellular technology.

Finally, the last chapter in the book, Chapter 25, "Wireless Use in Satellites and Space," takes you on a journey high above the earth, so that you can see some of the many ways that wireless technologies are used in satellite communications and in space research. You'll see how satellites can pinpoint any person's location on earth using Global Positioning System (GPS) technology. You'll also see how remarkable satellite telephones work. These telephones can make and receive calls anywhere on earth.

CHAPTER

22

How mCommerce and Corporate Wireless Access Work

ALTHOUGH wireless technology is used for entertainment, games, and to keep in touch with friends and family, one of its primary uses will be for business. There are some very obvious ways it is already used for business, primarily by allowing workers and managers to be easily reached, no matter where they are, via cell phones. And by extending the reach of corporate wireless networks, it is used by corporations as well.

But there are more important ways that wireless technologies will be used for business. One of the most important ones is called *mCommerce*, which stands for Mobile Commerce.

mCommerce refers to the use of cell phones or wireless personal digital assistants (PDAs) to do buying, primarily over the Internet. Although today there is very little mCommerce, its use is expected to explode in the future, in the same way that online shopping expanded in a few short years. Consider this: There are estimates that by 2003, the number of cell phone users worldwide will exceed the number of people who have fixed Internet access through phone lines, corporate networks, and devices such as cable modems. Given the chance, those cell phone users will want to shop—so much so that industry experts predict that by 2005, several billion dollars a year will be spent on mCommerce.

When we refer to mCommerce here, we don't refer to someone making a cell phone call and placing an order by voice. Instead, it means buying using the cell phone over the Internet, sometimes by browsing a wireless online shopping site and other times by making a direct connection to an individual store.

There's no single standard yet on how mCommerce will work, and mCommerce still is in its infancy, so it might be a while before everyone agrees on how it will work. But an emerging standard supported by several cell phone companies—Mobile Electronic Transactions (MET)—eventually might form the core of mCommerce. The standard has information for many different kinds of mCommerce, including buying directly from Web sites using the Wireless Applications Protocol (WAP), as well as for things such as wireless wallets, paying in retail stores using a cell phone, and other commercial applications.

Another important business use of wireless devices will be in allowing people with cell phones and PDAs to directly interact with large company databases and other "enterprise" software, no matter where they are. And it also will allow people to synchronize the data on their PDAs with corporate information, by updating the PDA information as well as updating the information found on corporate databases. This will be most important for sales people, but eventually many kinds of workers will be able to access corporate information this way. In fact, companies are already doing it, and it's a way of life at some corporations right now.

How mCommerce Works

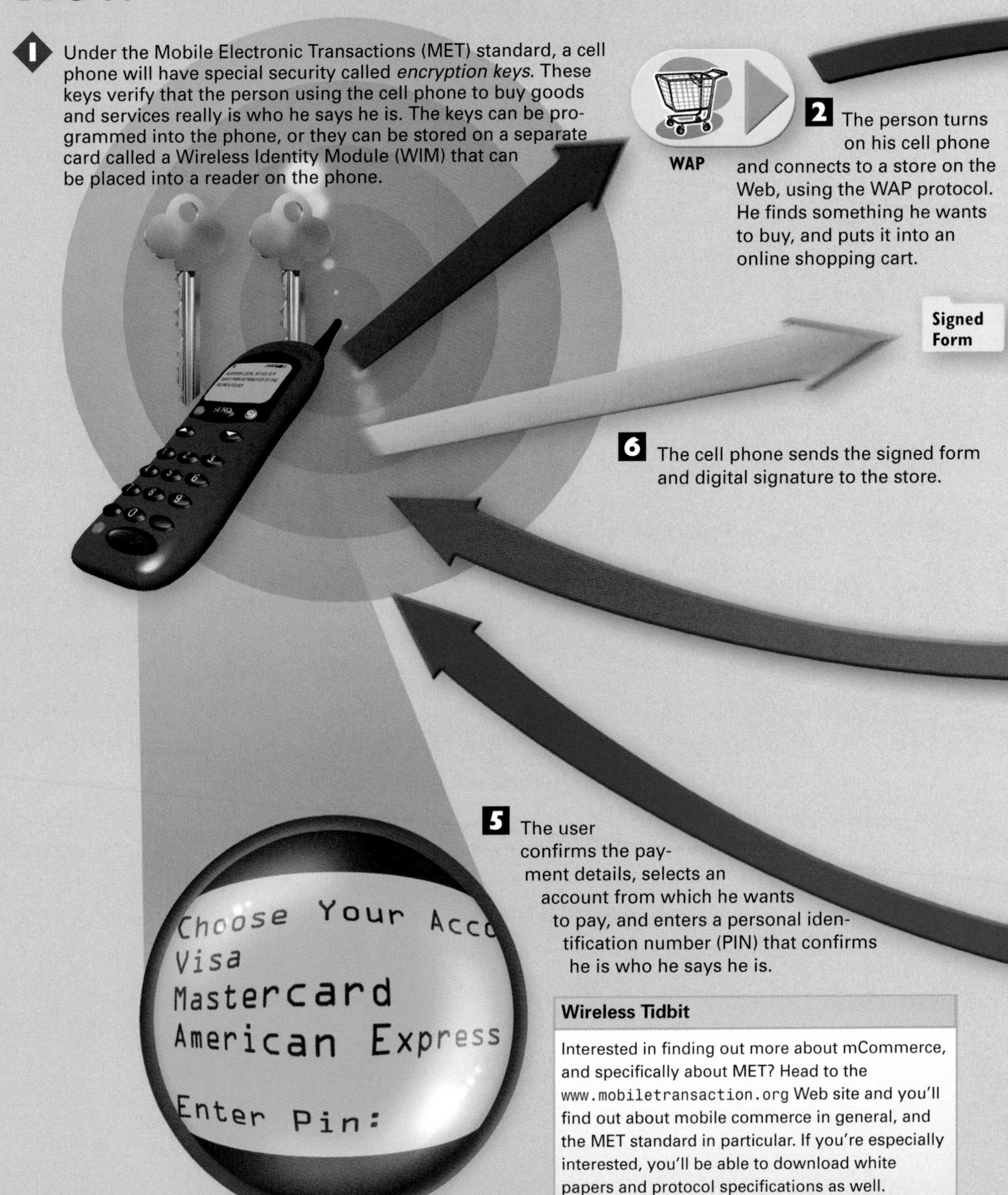

1 Under the Mobile Electronic Transactions (MET) standard, a cell phone will have special security called *encryption keys*. These keys verify that the person using the cell phone to buy goods and services really is who he says he is. The keys can be programmed into the phone, or they can be stored on a separate card called a Wireless Identity Module (WIM) that can be placed into a reader on the phone.

2 The person turns on his cell phone and connects to a store on the Web, using the WAP protocol. He finds something he wants to buy, and puts it into an online shopping cart.

6 The cell phone sends the signed form and digital signature to the store.

5 The user confirms the payment details, selects an account from which he wants to pay, and enters a personal identification number (PIN) that confirms he is who he says he is.

Wireless Tidbit

Interested in finding out more about mCommerce, and specifically about MET? Head to the `www.mobiletransaction.org` Web site and you'll find out about mobile commerce in general, and the MET standard in particular. If you're especially interested, you'll be able to download white papers and protocol specifications as well.

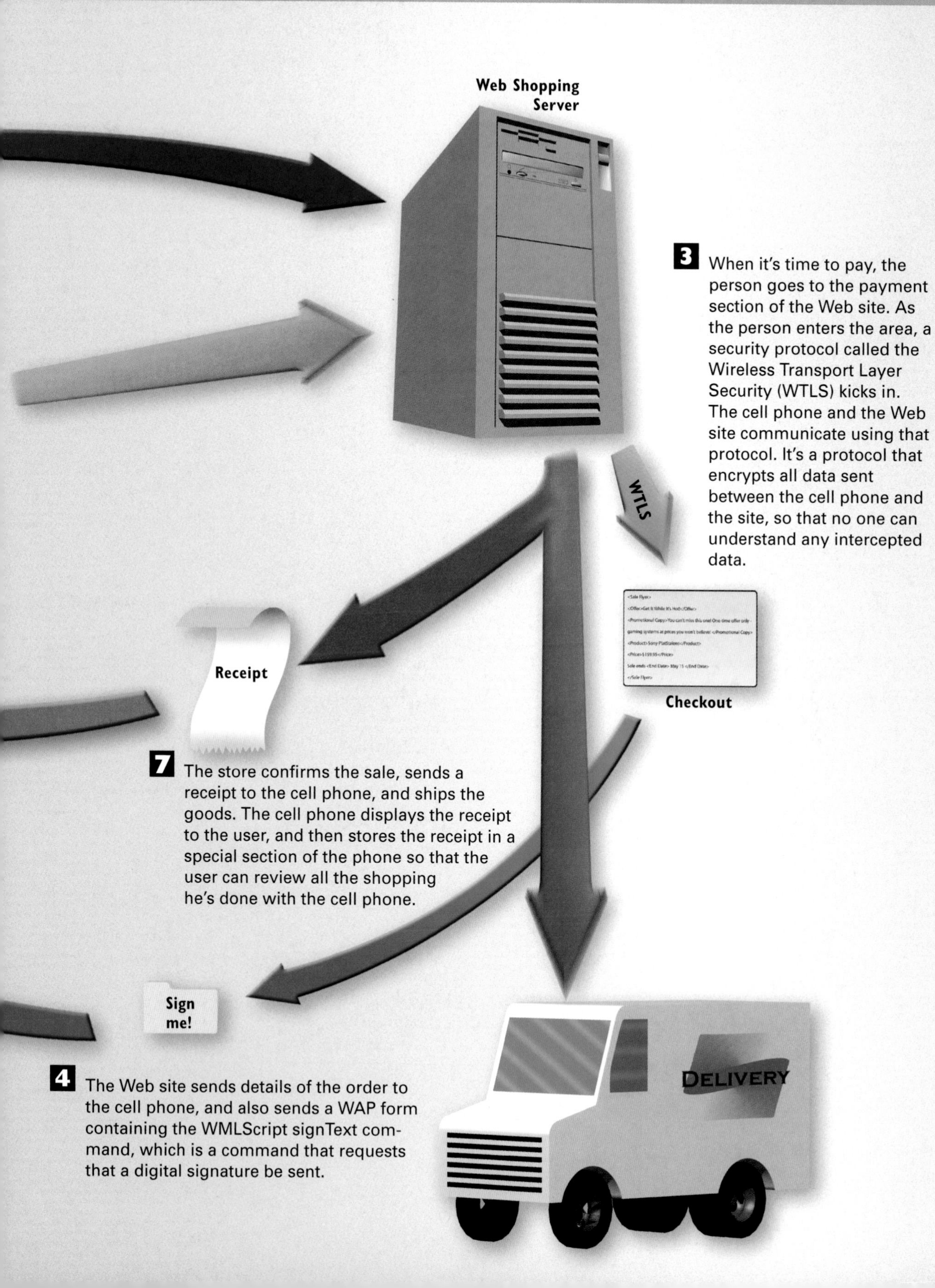

3 When it's time to pay, the person goes to the payment section of the Web site. As the person enters the area, a security protocol called the Wireless Transport Layer Security (WTLS) kicks in. The cell phone and the Web site communicate using that protocol. It's a protocol that encrypts all data sent between the cell phone and the site, so that no one can understand any intercepted data.

7 The store confirms the sale, sends a receipt to the cell phone, and ships the goods. The cell phone displays the receipt to the user, and then stores the receipt in a special section of the phone so that the user can review all the shopping he's done with the cell phone.

4 The Web site sends details of the order to the cell phone, and also sends a WAP form containing the WMLScript signText command, which is a command that requests that a digital signature be sent.

How Enterprise Systems Use Wireless Access

1 One of the most important uses of wireless technologies is for companies to give their workers and managers access to corporate computing resources no matter where they are. Personal digital assistants, such as the Palm and PocketPC, are used primarily for this, although as cell phones are given more computer-like functions, they can be used for this as well. In this scenario, before a person leaves the office, they have special software put onto their device that can synchronize its data with the corporation's large computers, such as large mainframe computers or servers. The person goes out on a sales call and brings along the wireless PDA.

5 The customer wants to place an order. The salesperson inputs the order directly into the PDA and leaves.

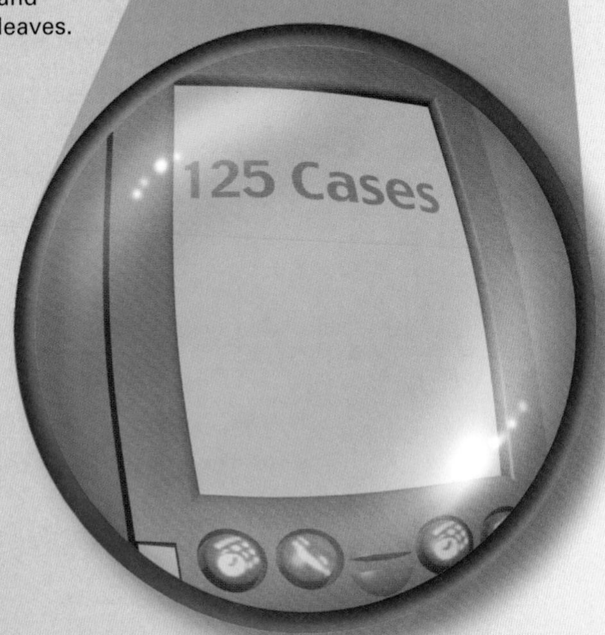

6 The salesperson makes several more calls throughout the day. Each time, he inputs the order directly into the PDA.

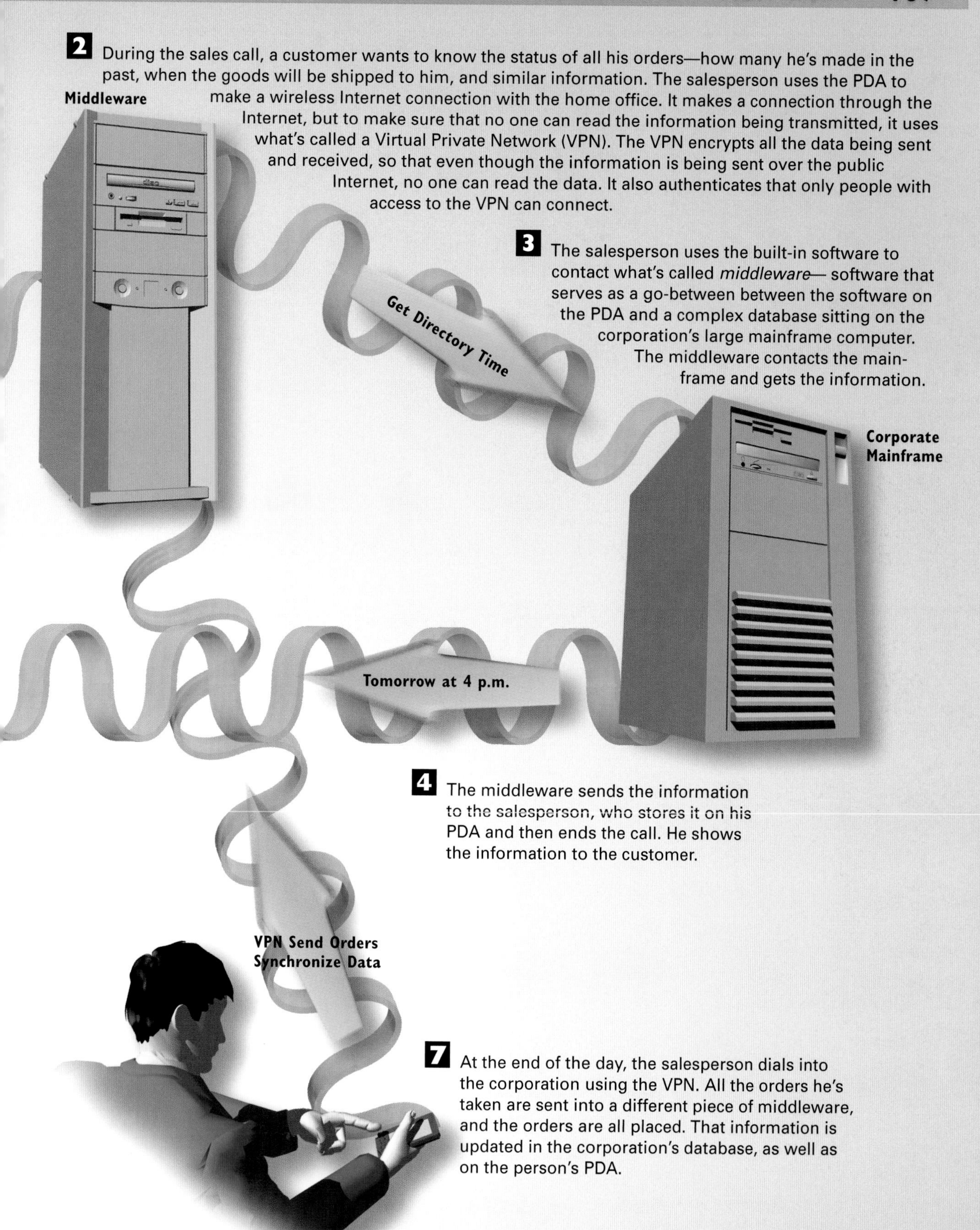

2 During the sales call, a customer wants to know the status of all his orders—how many he's made in the past, when the goods will be shipped to him, and similar information. The salesperson uses the PDA to make a wireless Internet connection with the home office. It makes a connection through the Internet, but to make sure that no one can read the information being transmitted, it uses what's called a Virtual Private Network (VPN). The VPN encrypts all the data being sent and received, so that even though the information is being sent over the public Internet, no one can read the data. It also authenticates that only people with access to the VPN can connect.

3 The salesperson uses the built-in software to contact what's called *middleware*— software that serves as a go-between between the software on the PDA and a complex database sitting on the corporation's large mainframe computer. The middleware contacts the mainframe and gets the information.

4 The middleware sends the information to the salesperson, who stores it on his PDA and then ends the call. He shows the information to the customer.

7 At the end of the day, the salesperson dials into the corporation using the VPN. All the orders he's taken are sent into a different piece of middleware, and the orders are all placed. That information is updated in the corporation's database, as well as on the person's PDA.

CHAPTER

23

Privacy and Security in a Wireless World

PICK up the newspaper on just about any given week, and you'll find scare stories about security and privacy problems having to do with the Internet. Viruses spreading worldwide, Web sites being hacked and attacked, people's identities stolen—these are just a few of the problems you'll find on the Internet. As of yet, you rarely hear similar scare stories when it comes to the wireless world. That might make you think that you're safe and secure, that hackers, pranksters and evil-doers want nothing to do with the wireless world.

Nothing could be further from the truth. Just wait. They're coming. In fact, ultimately, the wireless world might be more vulnerable than the Internet when it comes to privacy and security problems.

There are some simple reasons for that. The first is obvious: When you communicate wirelessly, you're sending information out through the air, so people can try to pluck that information using devices such as scanners, which can listen in on wireless communications. Thousands of hobbyists have long listened in on wireless police and emergency communications in this way—and they've also listened in on cell phone calls using scanners as well. Just think of the scandal a few years back when Prince Charles was overheard and taped talking to his mistress over a cell phone. And scanners also allow thieves to "clone" cell phones and allow other people to make free phone calls using that cloned phone.

Another reason for the vulnerability has to do with advances being made in cellular technology. Telephones increasingly are taking on the functions of computers, complete with computerized address books and databases and more features as well. The more complex telephones become, the more vulnerable they are to viruses and hackers.

Yet one more reason has to do with the always-on future of cellular communications. When it comes to computers, viruses are frequently sent through e-mail. In a world where cellular phones maintain an always-on connection to the Internet, e-mail, and chat programs, viruses can be sent instantaneously.

Not just cell phones are vulnerable. Personal digital assistants (PDAs), such as the Palm device, are targets as well. In fact, the Palm has already been hit by a cellular-borne virus that piggybacked a ride whenever a Palm beamed information to another Palm using infrared ports.

For corporations, the problems are even more serious. Wireless networks can be easily tapped by someone in a nearby parking lot using a laptop computer and inexpensive hardware and software. In fact, that's already happened. Two security experts have been making the rounds in Silicon Valley, listening in on wireless networks in many high-tech companies, including the hardware networking giant Sun Microsystems, and Nortel Networks, a company that specifically sells software designed to stop people from snooping in on networks.

So, the headlines might not be there today, but count on it—in the future, there will be bigger security dangers in the wireless world than there are today on the Internet.

What Dangers Are There to Privacy and Security?

Viruses can be spread wirelessly, so they pose potential dangers not just to computers, but also to cell phones, personal digital assistants (PDAs), and wireless networks. The viruses can be as innocuous as a joking text message on a cell phone or as damaging as deleting all the data from a PDA or crashing a wireless network.

Wireless network snoopers can see every bit of data traveling through wireless corporate networks and use that information to gain a competitive advantage on their competitors. Wireless hackers also can steal information, delete files, and destroy software.

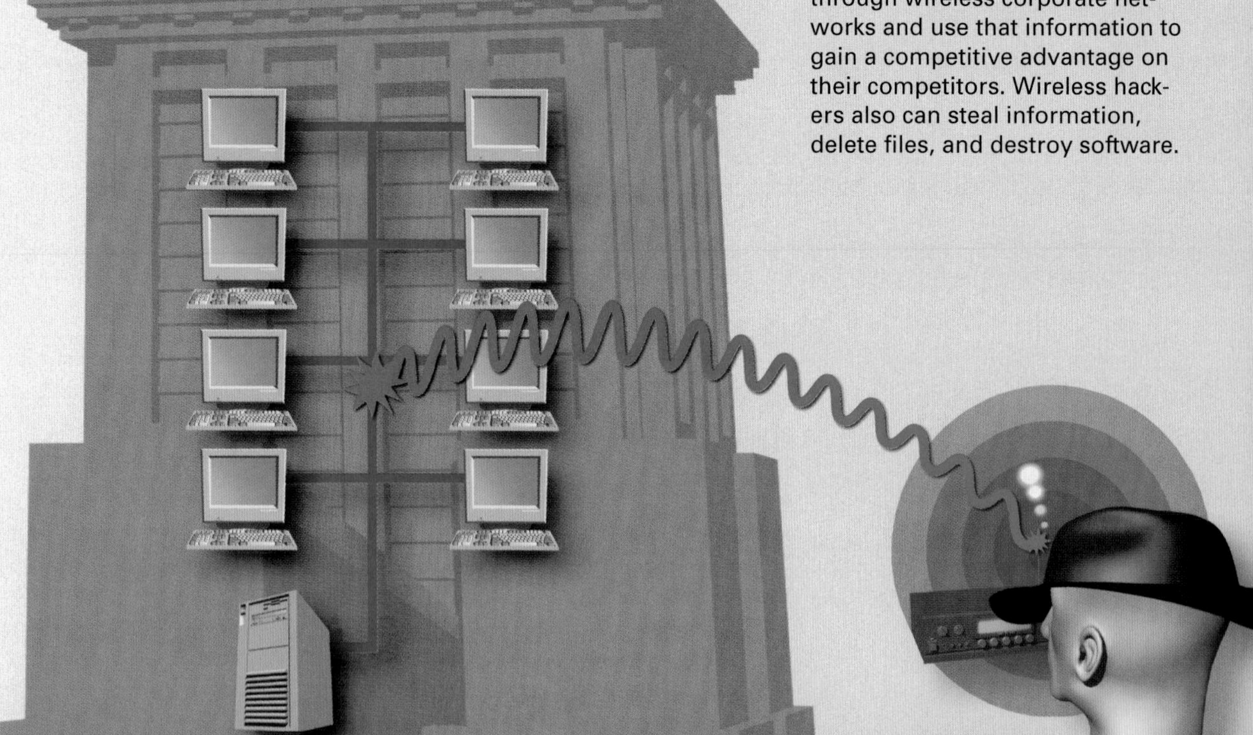

Cell phone snoopers can listen in on cell phone calls and invade people's and companies' privacy.

Cell phone cloners can steal the "identity" of cell phones and use that identity to make phone calls—leaving the owner of the cell phone holding the bill.

Wireless vandals can crash wireless networks by flooding them with phony information and messages so that the networks can't keep up with all the traffic.

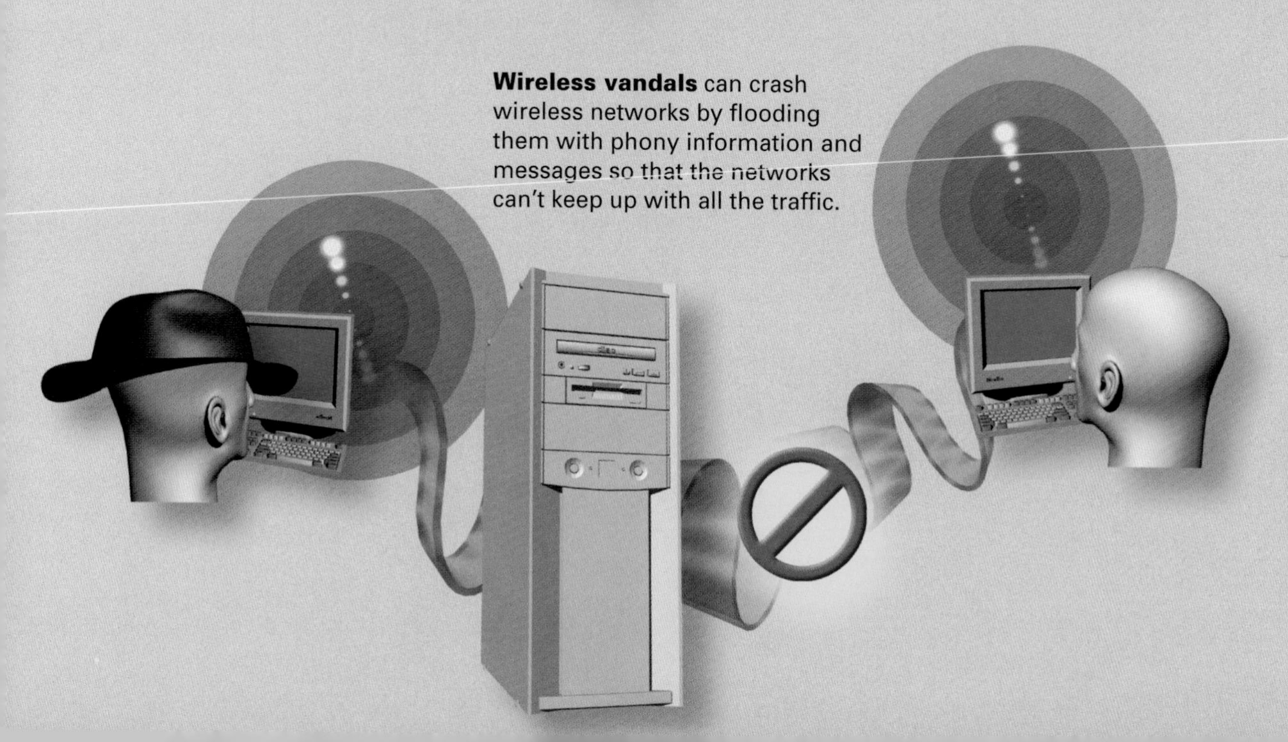

How Wireless Viruses Work

1 As of yet, there have not been many wireless viruses, but all that will change in coming years. The kinds of viruses vary according to the kind of wireless device being attacked. The first virus attack on cell phones, for example, was called Timofonica. It actually was a virus that used a computer to stage a prank attack on cell phones that used the Spanish company Telefonica cell phone service. In the first stage of the attack, people who used the Microsoft Outlook e-mail program each received an e-mail that had a file with a virus in it.

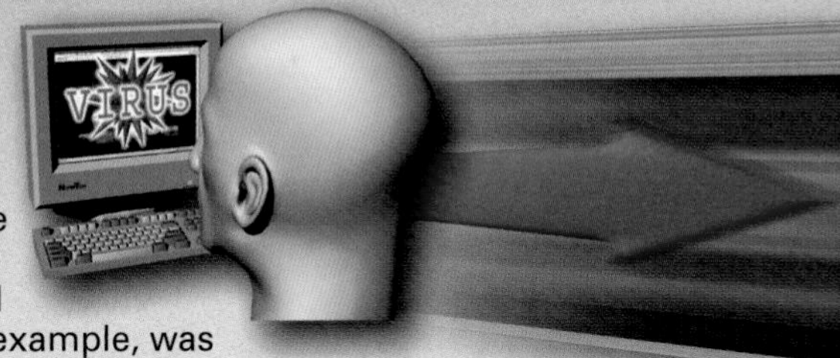

2 People then opened the file on their computers. The only people infected were those who opened the file. If the file wasn't opened, the virus couldn't infect anyone.

3 Opening the file turned the virus loose. The virus replicated itself on the person's computer, and sent copies of itself to every contact in the person's Outlook e-mail address book.

4 In addition to copying itself and sending copies to every contact in the address book, it also sent e-mail messages to a *Short Message Service (SMS)* gateway, directed at specific telephones of subscribers to the Telefonica service. SMS allows people on cell phones to receive text messages, similar to instant messaging on computers.

6 Telefonica subscribers began receiving the SMS messages generated by the virus. The messages did no harm to the cell phones, but they did clog the Telefonica cellular network. Although no harm was done in this instance, many people believe that in the future, cell phones will be targeted by real viruses, not just prank messages, especially when cell phones take on computer-like features and functions. Those viruses could be able to do things such as shut down the phone at random times or delete all the entries in a cell phone address book.

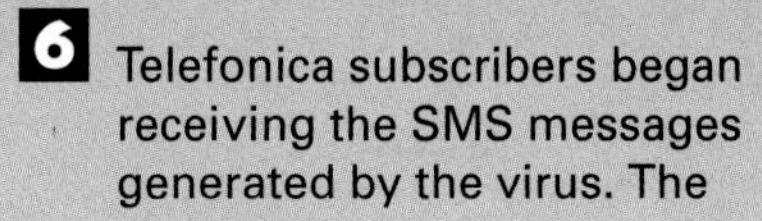

Gotcha Sucker!

5 The gateway converted the e-mail to a text message in SMS format, and sent the SMS message wirelessly.

Wireless Tidbit

Wireless viruses also have attacked the Palm PDA. In one instance, a virus called Phage was sent from Palm to Palm when the Palms exchanged data wirelessly through their infrared ports. When activated, the virus blanked the screen and halted the current program being run.

How Cell Phone Calls Can Be Tapped

1 Cell phone conversations can be listened to by a device called a *scanner*—the same device that enables people to listen in on police and emergency transmissions. Scanners can tune in to specific frequencies to see whether any transmissions are taking place. They also can automatically scan many frequencies, looking for transmissions. When they find transmissions, they can listen in on them.

2 The scanner can listen to the frequency the cell phone is using to transmit. But if it does that, it can hear only one side of the conversation—the voice of the person using the cell phone, not the person to whom he's talking.

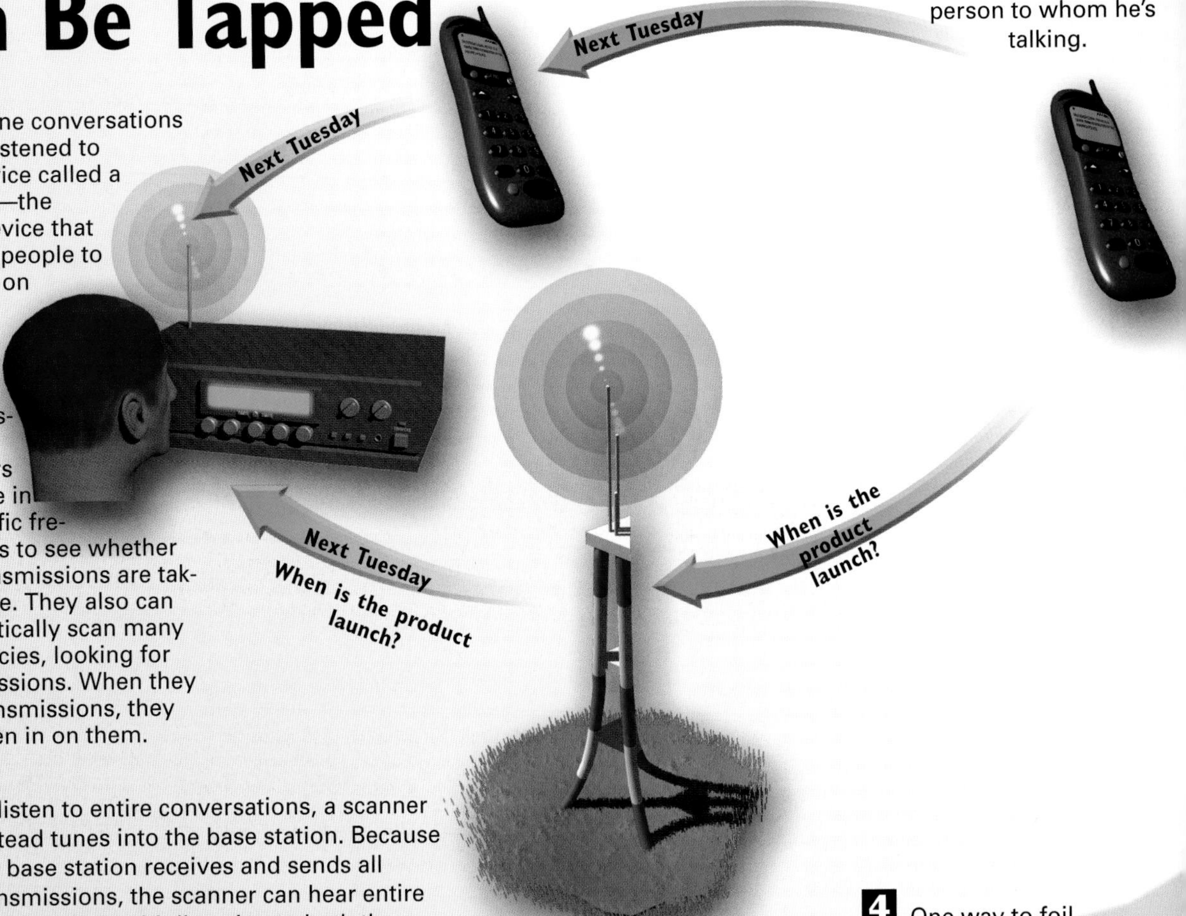

3 To listen to entire conversations, a scanner instead tunes into the base station. Because the base station receives and sends all transmissions, the scanner can hear entire conversations—it's listening to both the receiving and transmitting frequencies simultaneously, referred to as *full duplex*.

4 One way to foil scanners is to use digital technology. With digital technology, conversations can be *encrypted*—scrambled so that if someone listens in using a scanner, they won't be able to understand what's being said. That's why digital phones are more secure than analog ones.

Next Tuesday
Next Tuesday
When is the product launch?
Next Tuesday
When is the product launch?
Next Tuesday

Wireless Tidbit

If a scanner is listening in on a conversation of someone driving a car, or otherwise moving from location to location, the scanner can easily lose the transmission. That's because when someone is moving, they'll often move from one cell to another, and are handed off from the base station of one cell to the base station of another cell. When that handoff happens, the scanner loses the transmission because it was tuned in to the base station of the first cell.

5 Cell phones aren't the only wireless communications that can be tapped. Using a laptop computer and inexpensive scanning hardware and software, someone could listen in on an entire wireless network and capture all the data and messages being sent across it.

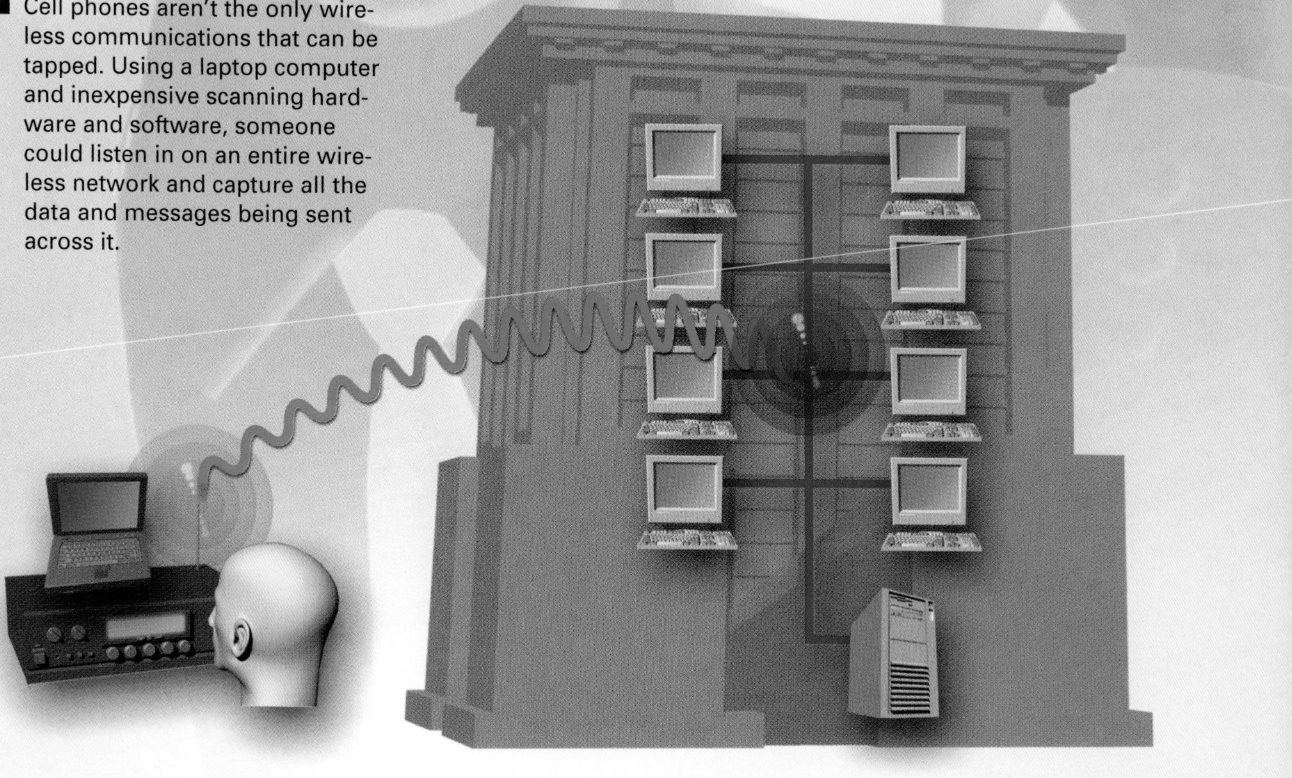

How Cell Phone Identities Can Be Stolen

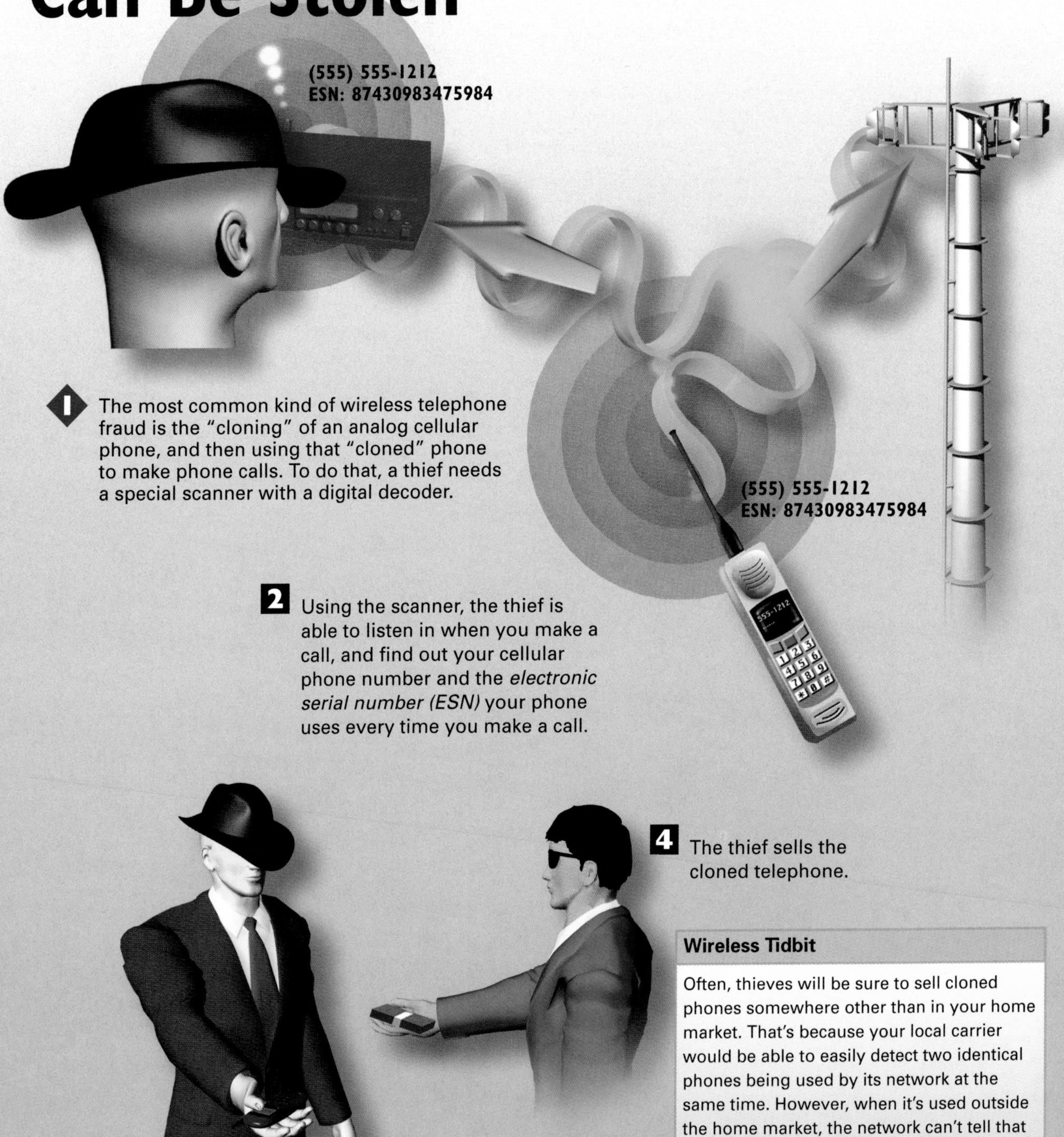

1 The most common kind of wireless telephone fraud is the "cloning" of an analog cellular phone, and then using that "cloned" phone to make phone calls. To do that, a thief needs a special scanner with a digital decoder.

2 Using the scanner, the thief is able to listen in when you make a call, and find out your cellular phone number and the *electronic serial number (ESN)* your phone uses every time you make a call.

4 The thief sells the cloned telephone.

Wireless Tidbit

Often, thieves will be sure to sell cloned phones somewhere other than in your home market. That's because your local carrier would be able to easily detect two identical phones being used by its network at the same time. However, when it's used outside the home market, the network can't tell that two identical phones are being used.

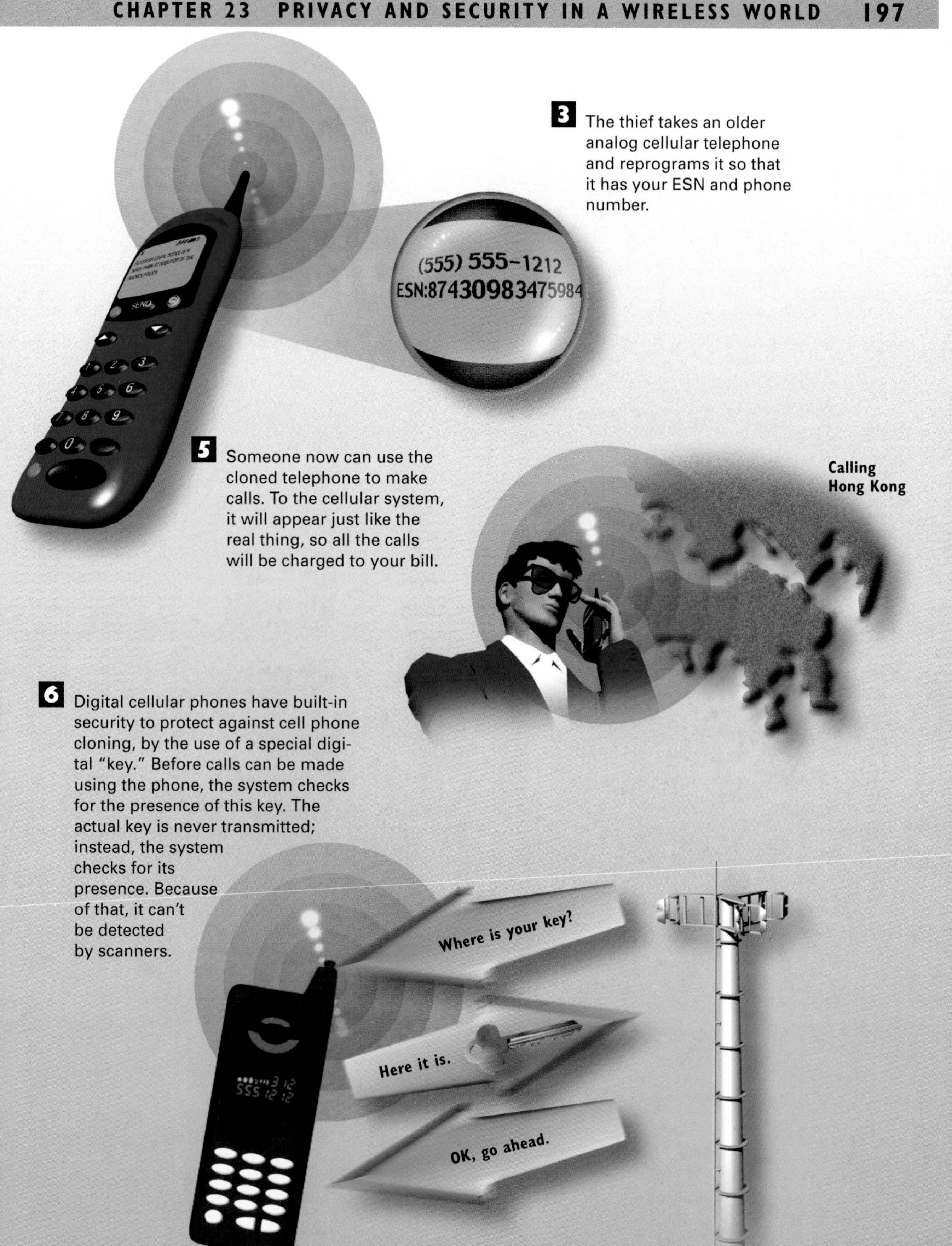
(555) 555–1212
ESN:87430983475984
SEND
Calling
Hong Kong
Where is your key?
Here it is.
OK, go ahead.

3 The thief takes an older analog cellular telephone and reprograms it so that it has your ESN and phone number.

5 Someone now can use the cloned telephone to make calls. To the cellular system, it will appear just like the real thing, so all the calls will be charged to your bill.

6 Digital cellular phones have built-in security to protect against cell phone cloning, by the use of a special digital "key." Before calls can be made using the phone, the system checks for the presence of this key. The actual key is never transmitted; instead, the system checks for its presence. Because of that, it can't be detected by scanners.

CHAPTER

24 How Wireless 3G Works

THE services you can get with today's cell phone will look nothing like the ones you'll get tomorrow. They'll be so dramatically different, it almost will be as if you're using a different device altogether—and in fact, you will be.

Think of today's cellular phone systems as the Internet before the invention of the World Wide Web. Before the Web, the Internet largely was a text-based system; a sleepy kind of backwater. Then the Web came, and the world changed.

In the same way, a group of cellular technologies, collectively known as 3G, for *third generation*, will forever change the way in which cell phones and similar devices are used. These technologies will offer always-on, high-speed cellular access to the Internet as well as to phone calls. It's hard to know exactly how the world will change because of it, but there's no doubt that it will. You'll receive video and music on your cell phone, get live navigational directions, receive instant e-mail, and, no doubt, a lot more, as well.

3G is not a single, distinct, technology, but rather a catchall phrase that encompasses a group of technologies to offer this always-on, high-speed access. The technologies are a natural outgrowth of previous cellular technologies. So-called *1G* technologies were the initial wave of cell phones and wireless devices. It's hard to remember now, but these first cell phones and pagers were heavy and bulky and didn't offer that many services—cell phones were attached to backpacks that had batteries in them; pagers were brick-sized.

2G technologies refer to the technology we have today—smaller-size cell phones and pagers; simple cellular Web browsing and e-mail delivery; and other technologies, such as caller ID. Some people even talk about 2.5G technologies, which to a great extent means today's technology, but delivered at a higher speed. In fact, some people refer to GPRS technology as a 2.5G rather than a 3G technology.

Billions of dollars have been spent worldwide on the necessary infrastructure for delivering 3G technologies. Most of that money has been spent in Japan and Europe, and that's where 3G technologies will first hit. The Japanese company NTT DoCoMo worked on rolling out some of the first 3G technologies in late 2001 and early 2002, with the world following from there.

Although there's no single standard that defines what 3G technology is, it's sometimes referred to as IMT-2000 (International Mobile Communications-2000). An international organization called the International Telecommunications Union (ITU) has put together a broad group of specifications known collectively as IMT-2000. But the truth is, at this point, it's little more than an alphabet soup of technologies that won't necessarily be the same from continent to continent.

How Wireless 3G Works

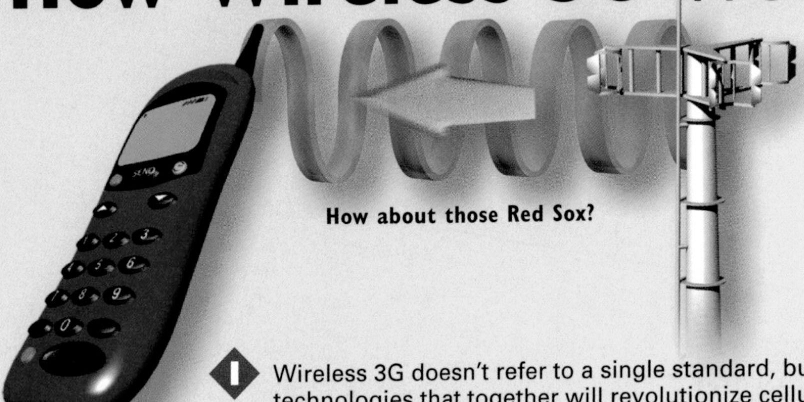

1. Wireless 3G doesn't refer to a single standard, but instead to a collection of technologies that together will revolutionize cellular telephones and meld them with the Internet. One of the most basic pieces of 3G is that your cellular connection will be always on—you won't have to dial to connect to the Internet. So, you'll be able to get instant e-mail the moment it's in your inbox, for example, or be able to do instant messaging without making a cell phone call.

3. One of the most important parts of 3G is that it will offer much higher-speed connections than are currently available. Because there is no single 3G standard, there's no definitive speed at which a service can be called 3G. However, 3G speeds probably will be at least 384 kilobits per second (kbps), and up to more than 2 megabits per second (Mbps). (As a practical matter, individual cell phones generally won't access the network at those speeds, because the data rates are shared among all users in a given cell.) This might allow for things such as video being delivered to cell phones. A variety of technologies can be used to deliver this higher speed, including Wideband Code Division Multiple Access (W-CDMA) and Code Division Multiple Access (CDMA). For more information about these technologies, turn to Chapter 11, "How Cellular Telephones Work."

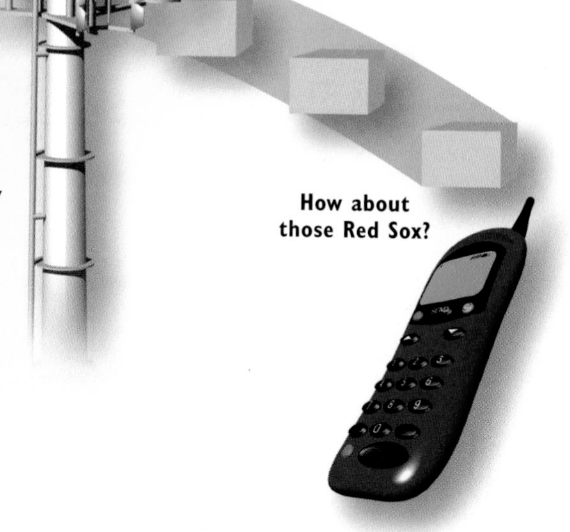

How about those Red Sox?

2 3G technologies will work more like the Internet than today's cellular telephones. They'll use *packet* technology, which means that, instead of sending information as single entire units, they'll break up information into smaller packets and send those packets individually, so they can be reassembled on the receiving end. One way they might do this is by using General Packet Radio Service (GPRS) technology.

How about those Red Sox?

Next right for Fenway Park

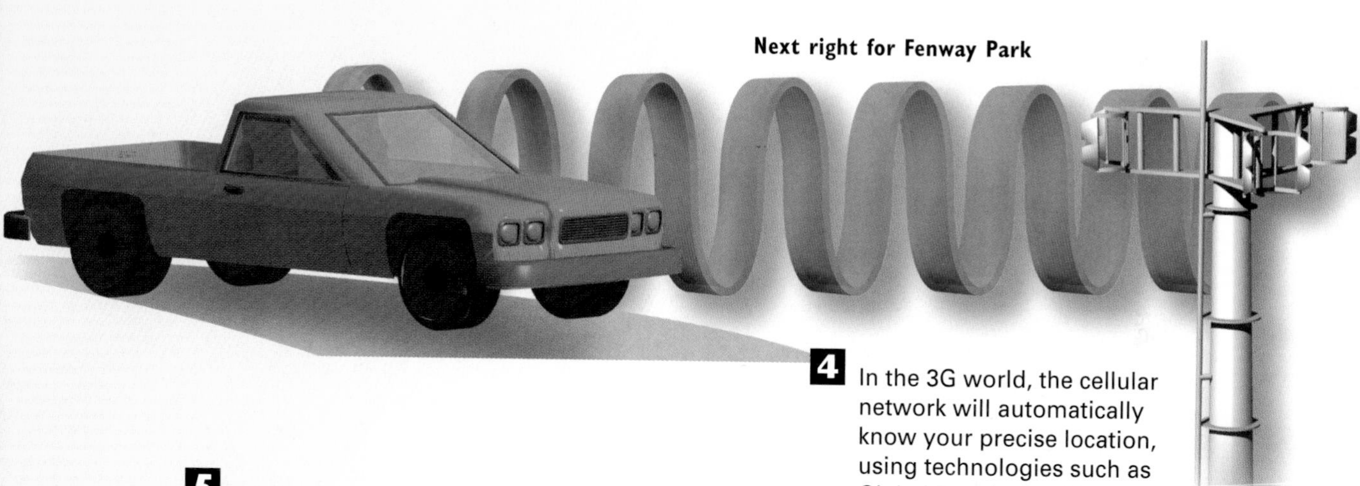

4 In the 3G world, the cellular network will automatically know your precise location, using technologies such as Global Positioning Satellites (GPS). Because of that, new services, such as automatic mapping, can be delivered to you.

5 All this will lead to the development of new services and to new ways in which cell phones can be used. You'll be able to watch videos on your cell phones, get MP3 files and other music files delivered to your phone, browse the Web at high speeds, do cell phone–based videoconferencing, get live navigational directions when you drive, and much more.

Web Server

CHAPTER

25 Wireless Use in Satellites and Space

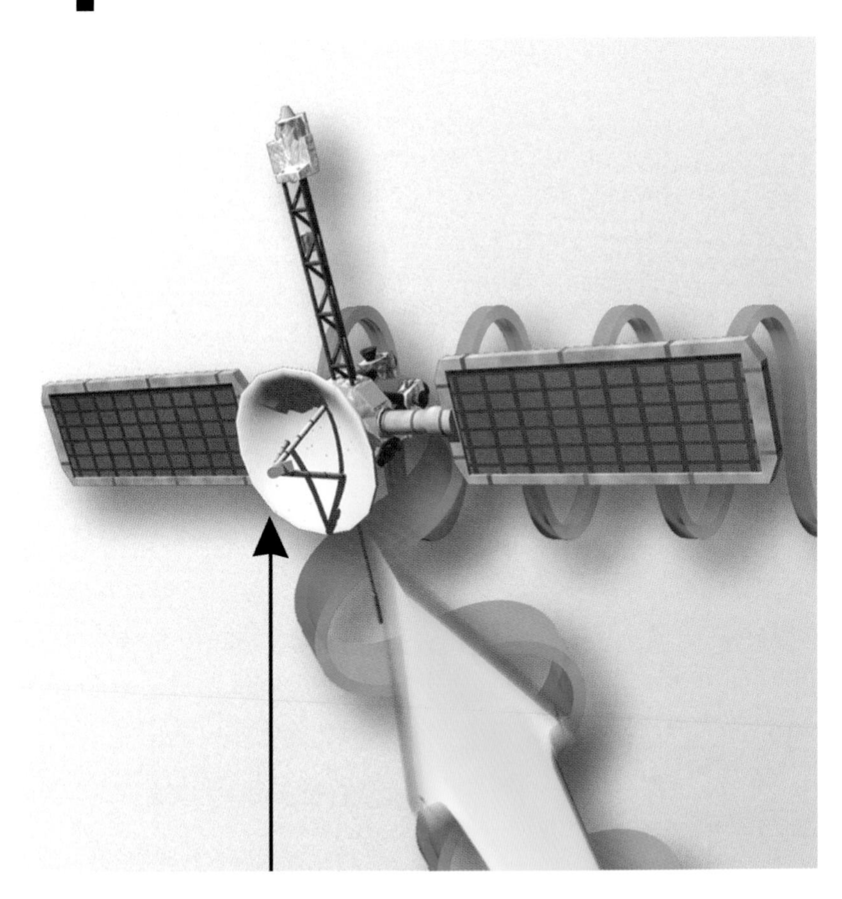

YOU might not realize it, but orbiting above the earth and you are hundreds of satellites busy beaming information to and from the earth. They're tracking the weather or gathering spy information; sending phone calls, TV, and radio signals; sending positioning information to people; and far more.

Some of them are sending information from far out in space—from Mars, for example, or even from beyond the limits of our solar system. And for all these communications, they're using wireless technologies. What might be most surprising about satellite communications is that they use the same basic kinds of communications equipment used here on earth—transmitters, antennas, receivers, and so on.

Several types of satellites are involved in communications. *Geostationary* satellites orbit the earth at 35,784 kilometers. At that height, they have an orbital period of 24 hours—in other words, they are orbiting at the same speed the earth revolves. So, these satellites can stay fixed above the exact same position above a spot on the earth. Because of this, satellite dishes on earth can simply point at one of these satellites and never need to move, because the relative position of the satellite doesn't move. They are located in a belt above the equator, so antennas in the northern hemisphere that use these satellites all point in a southern direction.

A second kind of satellite is *Middle Earth Orbit (MEO)* satellites, which orbit at an altitude of between 5,000 and 15,000 kilometers above the earth. Among these are satellites for the *Global Positioning System (GPS)*. As you'll see in the illustration later in this chapter, GPS systems can pinpoint your location on earth so that you can know your current longitude and latitude. When combined with computer technology and a database of maps, they can provide navigation instructions. They also can provide exceedingly accurate time.

A third kind of satellite are *Low Earth Orbit (LEO)* satellites, which orbit at a height of 100 to 1,000 kilometers. The very first communications satellite, the Echo satellites, were LEO satellites, launched in 1960. For financial and technical reasons, these kinds of satellites were used less and less until the 1990s, when a company called Iridium hatched a scheme to launch dozens of LEO satellites that would allow people to make and receive phone calls from anywhere on earth, by beaming calls to and from the satellites. The business went bankrupt in a spectacular fashion, but it has since been revived, and several other companies, such as Globalstar, now are planning to sell satellite phone service as well.

The other kinds of satellites that use wireless communications are those used for space exploration. When you see pictures from Mars or Jupiter, those photos have been beamed to earth using wireless technologies. Surprisingly, as you'll see in the illustration later in this chapter, the transmitters on them are not particularly strong—only about eight times as strong as those on a cell phone. But a variety of other technologies and designs make up for that, so that the satellites can keep sending data back to earth, literally from beyond the edge of the solar system.

How Global Positioning Satellites (GPS) Work

1 To understand how GPS works, you first need to understand the concept of *triangulation*, which allows you to determine exactly where you are if you know your distance from three different points. Let's say that you know you're 75 miles from Boston. You could be anywhere on this circle, which has Boston in its center.

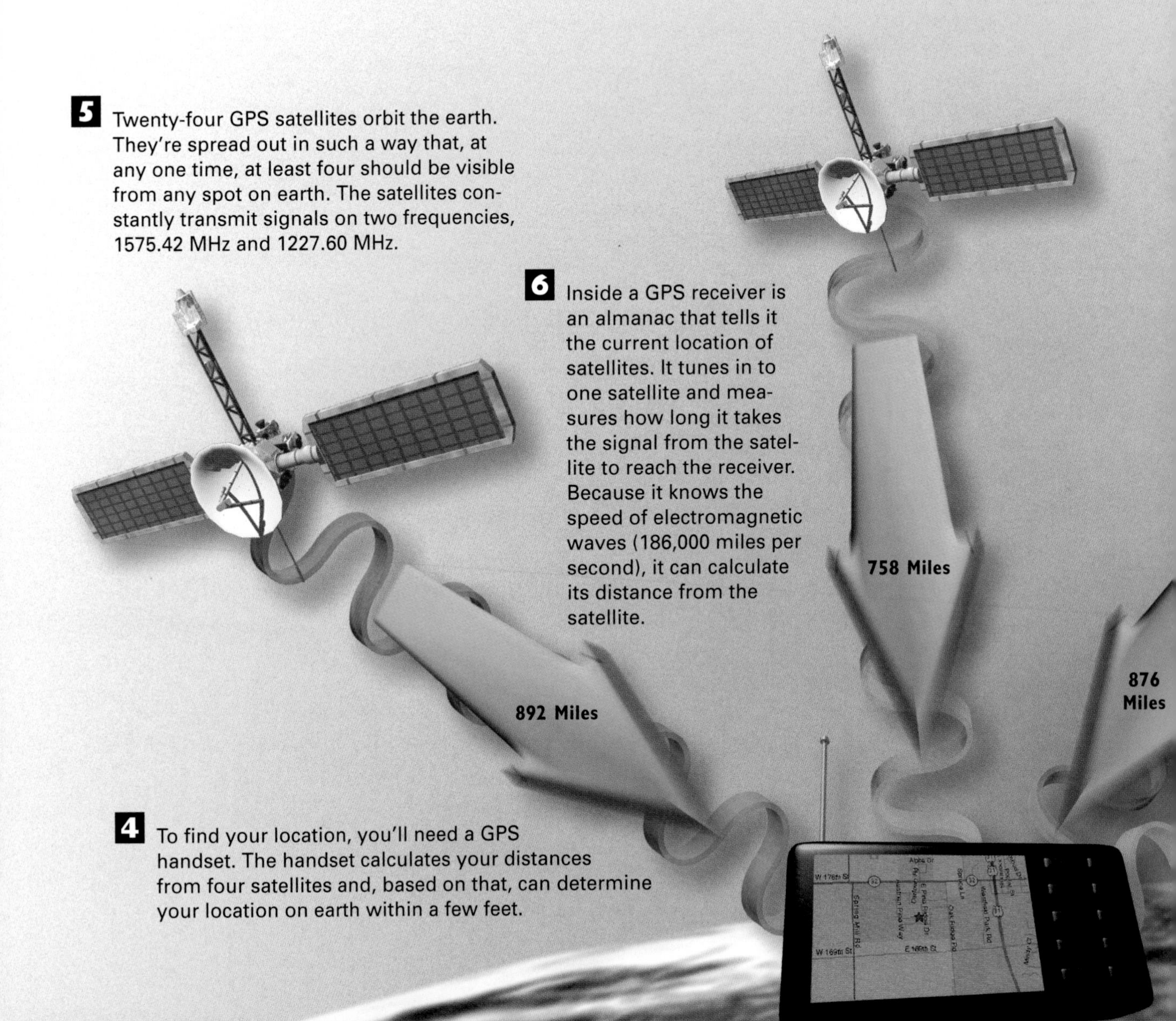

5 Twenty-four GPS satellites orbit the earth. They're spread out in such a way that, at any one time, at least four should be visible from any spot on earth. The satellites constantly transmit signals on two frequencies, 1575.42 MHz and 1227.60 MHz.

6 Inside a GPS receiver is an almanac that tells it the current location of satellites. It tunes in to one satellite and measures how long it takes the signal from the satellite to reach the receiver. Because it knows the speed of electromagnetic waves (186,000 miles per second), it can calculate its distance from the satellite.

4 To find your location, you'll need a GPS handset. The handset calculates your distances from four satellites and, based on that, can determine your location on earth within a few feet.

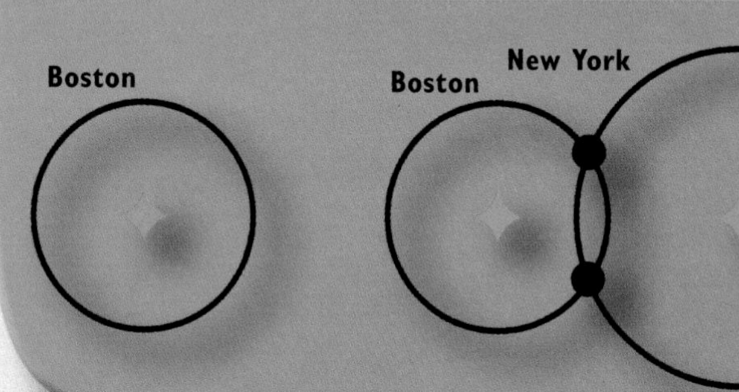

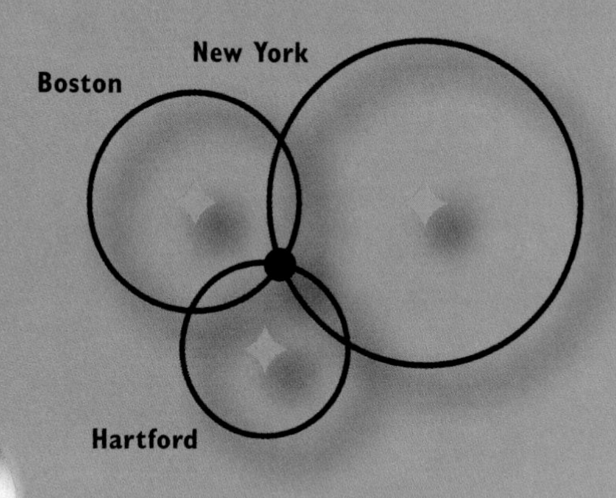

2 Now suppose you know that you're 170 miles from New York. Drawing a second circle with New York as its center, you now can be on either points A or B, where the two circles intersect.

3 Finally, you know your distance from a third point, Hartford. You're 20 miles from Hartford. You draw a third circle with Hartford in the middle, and you know your precise location. With GPS systems, though, you'll know your distance not from points on earth, but from satellites circling above it. So, when you know your distance from one satellite, you can be anywhere not on a circle around it, but in a sphere around it. The three spheres intersect at two points, so, theoretically, there are two possible points where you can be. However, one of those points is in space, so you can use distance measurements from three satellites to determine where you are. However, for greatest accuracy, and to get information such as your altitude, you need to measure your distance from four satellites.

Wireless Tidbit

The GPS system was developed not for civilian use, but instead by the U.S. military. The first GPS satellite was launched in 1978. The system became partially operational in 1986 and fully operational in 1990. The military still runs the system.

7 The receiver does the same thing with three more satellites. It now can tell you the longitude and latitude of where you are on earth.

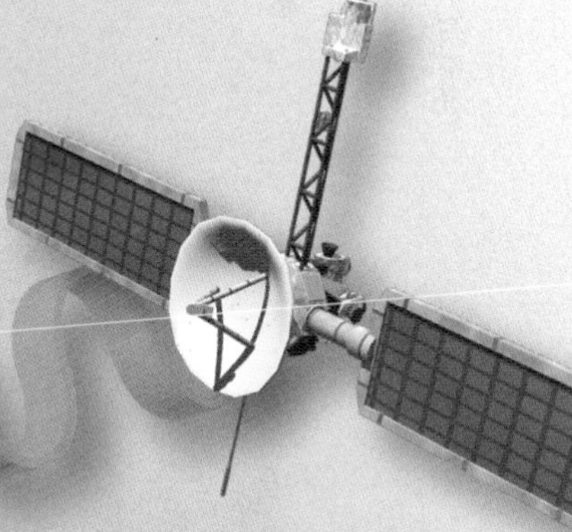

975 Miles

8 GPS receivers also can include maps, so you can see where you are on a map. And they can constantly track where you are, so they can be used in car navigation systems to show you where you are driving. When a GPS is combined with a map and database of streets and directions, a system can give you driving directions and change those directions as you drive.

How Satellite Phones Work

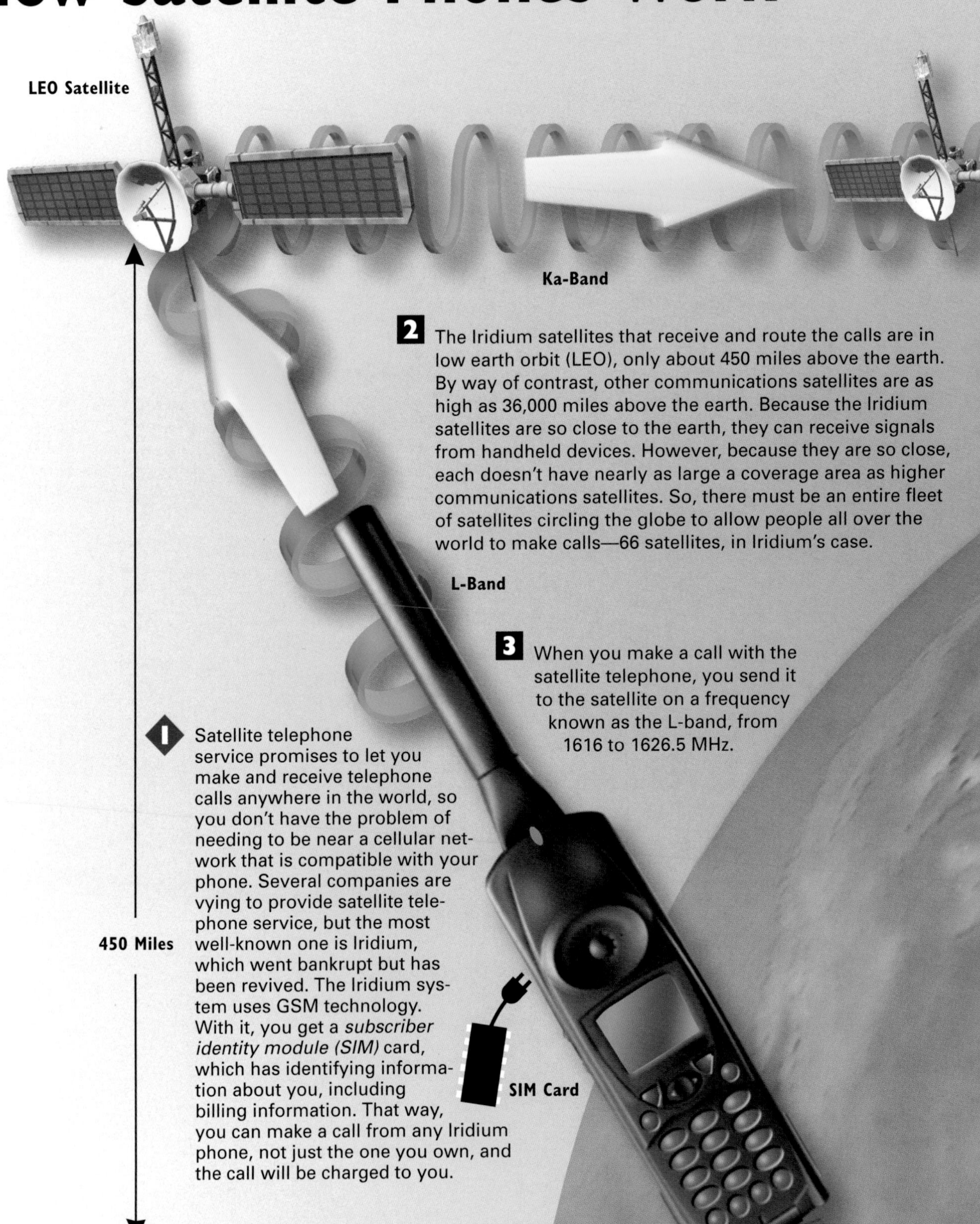

1 Satellite telephone service promises to let you make and receive telephone calls anywhere in the world, so you don't have the problem of needing to be near a cellular network that is compatible with your phone. Several companies are vying to provide satellite telephone service, but the most well-known one is Iridium, which went bankrupt but has been revived. The Iridium system uses GSM technology. With it, you get a *subscriber identity module (SIM)* card, which has identifying information about you, including billing information. That way, you can make a call from any Iridium phone, not just the one you own, and the call will be charged to you.

2 The Iridium satellites that receive and route the calls are in low earth orbit (LEO), only about 450 miles above the earth. By way of contrast, other communications satellites are as high as 36,000 miles above the earth. Because the Iridium satellites are so close to the earth, they can receive signals from handheld devices. However, because they are so close, each doesn't have nearly as large a coverage area as higher communications satellites. So, there must be an entire fleet of satellites circling the globe to allow people all over the world to make calls—66 satellites, in Iridium's case.

3 When you make a call with the satellite telephone, you send it to the satellite on a frequency known as the L-band, from 1616 to 1626.5 MHz.

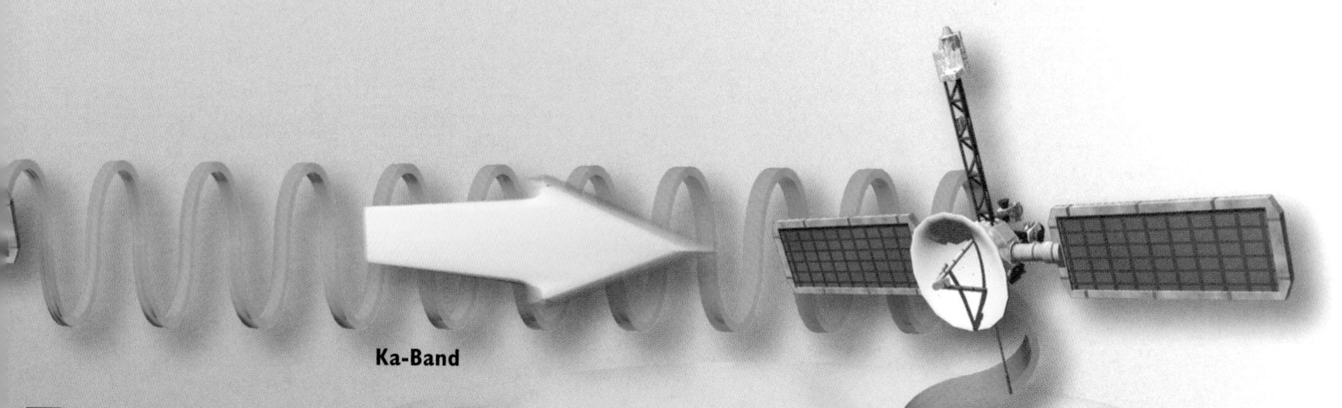

4 The satellite receives the call. In some other kinds of satellite systems, the call would now have to be beamed to earth and routed through terrestrial phone networks. But in Iridium's case, the satellites can work much like a cellular network—calls can be routed from satellite to satellite. A transponder on the satellite changes the frequency of the call from the L-band to what is known as the Ka band, at 23.18 to 23.38 GHz. The call then is sent to a satellite closer to the call's final destination.

5 The call is routed from satellite to satellite in this manner until it is near the call's final destination.

Wireless Tidbit

Wonder where the Iridium satellite network got its name? Believe it or not, from the periodic table of elements. Originally, the network was supposed to have 77 satellites in it. The 77th element in the periodic table is Iridium; hence, the company's name.

6 The call is beamed down from the satellite to the Iridium gateway closest to the call's destination, using the L-band.

7 The gateway has links to the public phone network and cellular networks, and sends the call to the proper network so that it can be completed.

How Space Exploration Satellites Use Wireless Communications

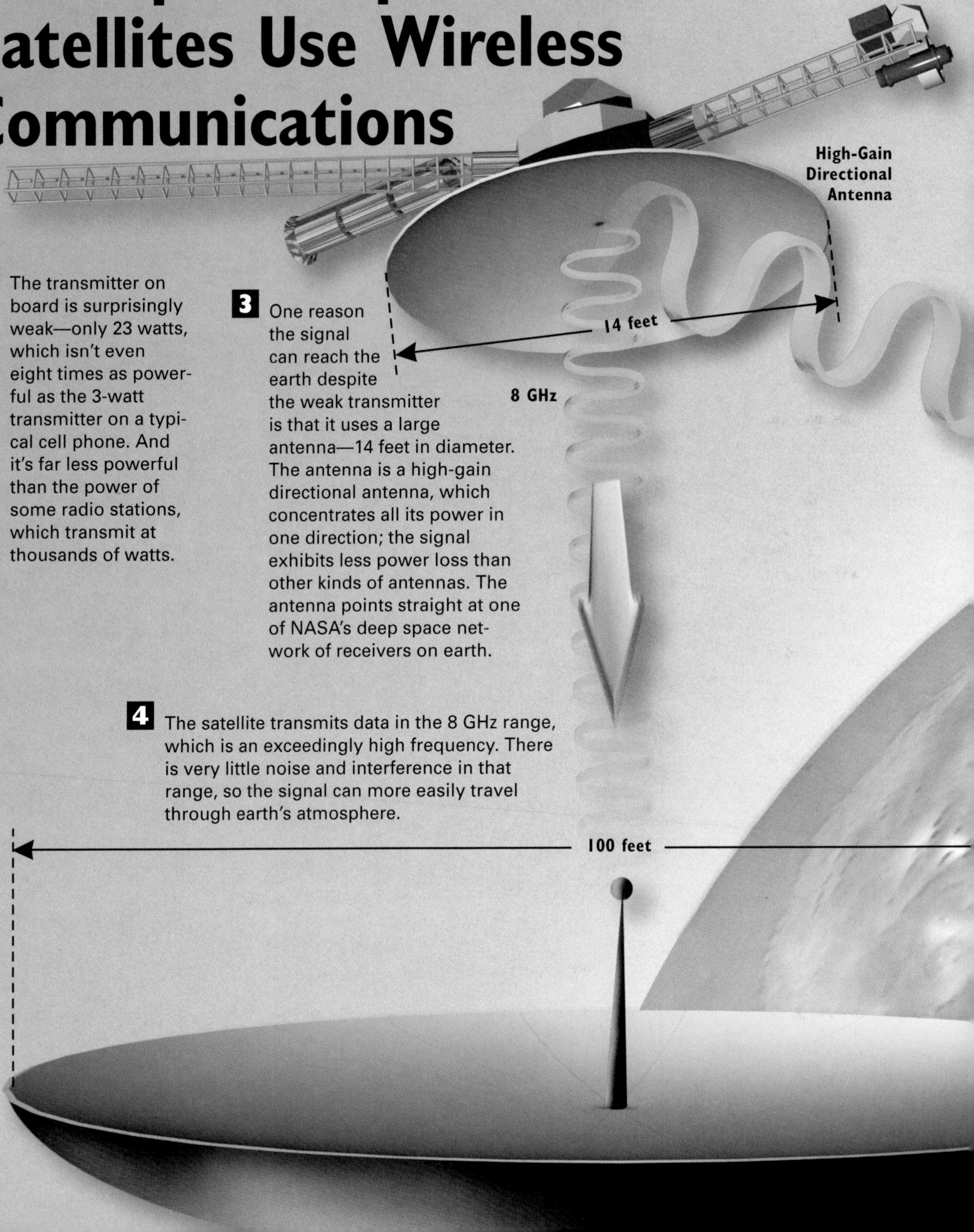

High-Gain Directional Antenna

14 feet

8 GHz

100 feet

2 The transmitter on board is surprisingly weak—only 23 watts, which isn't even eight times as powerful as the 3-watt transmitter on a typical cell phone. And it's far less powerful than the power of some radio stations, which transmit at thousands of watts.

3 One reason the signal can reach the earth despite the weak transmitter is that it uses a large antenna—14 feet in diameter. The antenna is a high-gain directional antenna, which concentrates all its power in one direction; the signal exhibits less power loss than other kinds of antennas. The antenna points straight at one of NASA's deep space network of receivers on earth.

4 The satellite transmits data in the 8 GHz range, which is an exceedingly high frequency. There is very little noise and interference in that range, so the signal can more easily travel through earth's atmosphere.

1 Space exploration satellites, such as the two Voyager satellites, use RF technology to send data and pictures back to earth and to receive instructions from earth.

Wireless Tidbit

Voyager 1 was launched on September 5, 1977. It passed by Jupiter in March 1979 and Saturn in November 1980, and then continued out of the solar system.

Voyager 2 was launched August 20, 1977. It passed Jupiter in July 1979, Saturn in August 1981, Uranus in January 1986, and Neptune in August 1989. They are both in deep space, traveling away from our solar system and sending back signals and data. They're expected to continue operating for 25 years, although at some point they will be too far away for us to receive the signals.

7 Transmitters that send instructions to the satellite are very powerful—tens of thousands of watts. They need to be so powerful because the satellite doesn't have a large sensitive antenna or a powerful amplifier onboard.

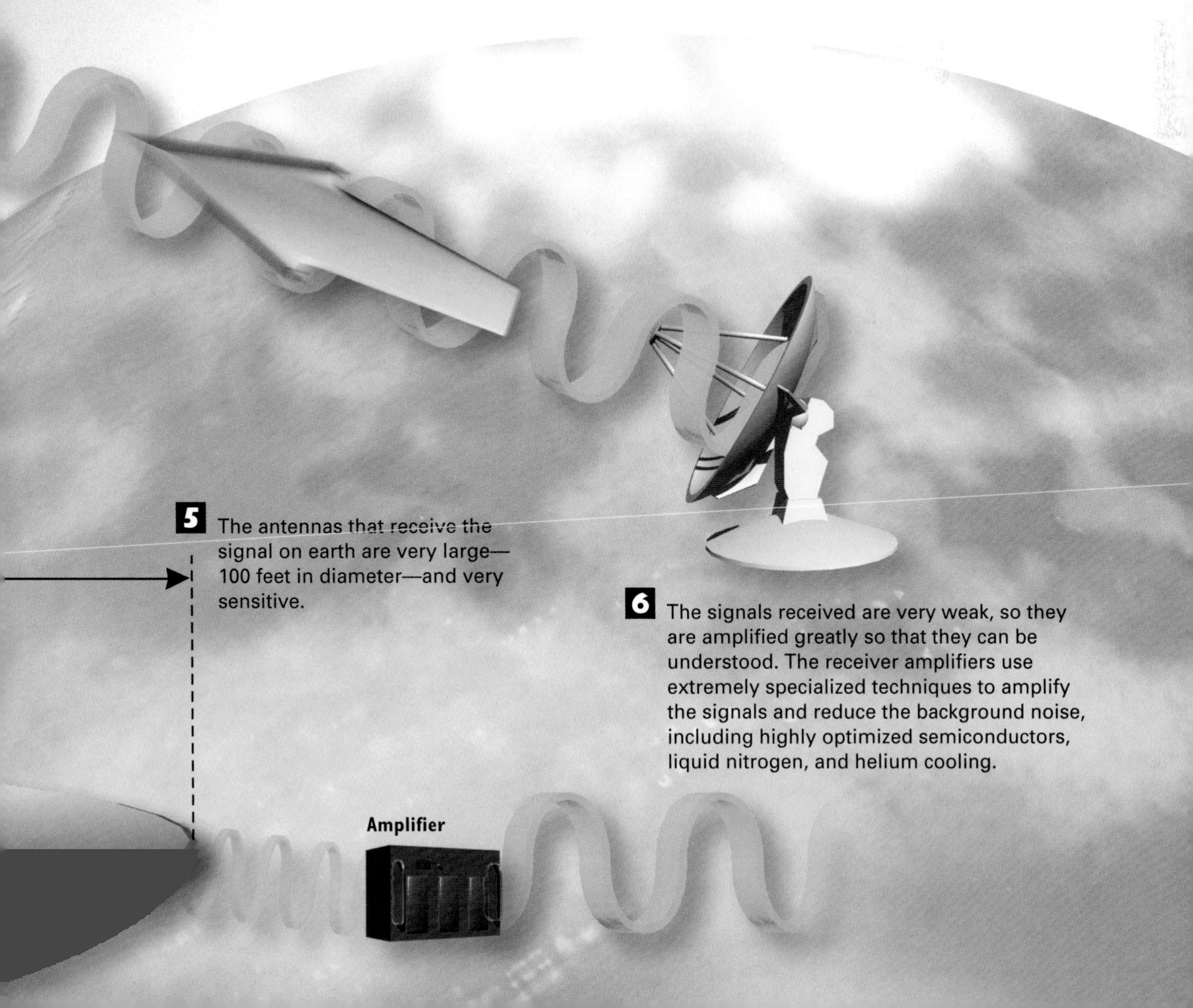

5 The antennas that receive the signal on earth are very large—100 feet in diameter—and very sensitive.

6 The signals received are very weak, so they are amplified greatly so that they can be understood. The receiver amplifiers use extremely specialized techniques to amplify the signals and reduce the background noise, including highly optimized semiconductors, liquid nitrogen, and helium cooling.

Glossary

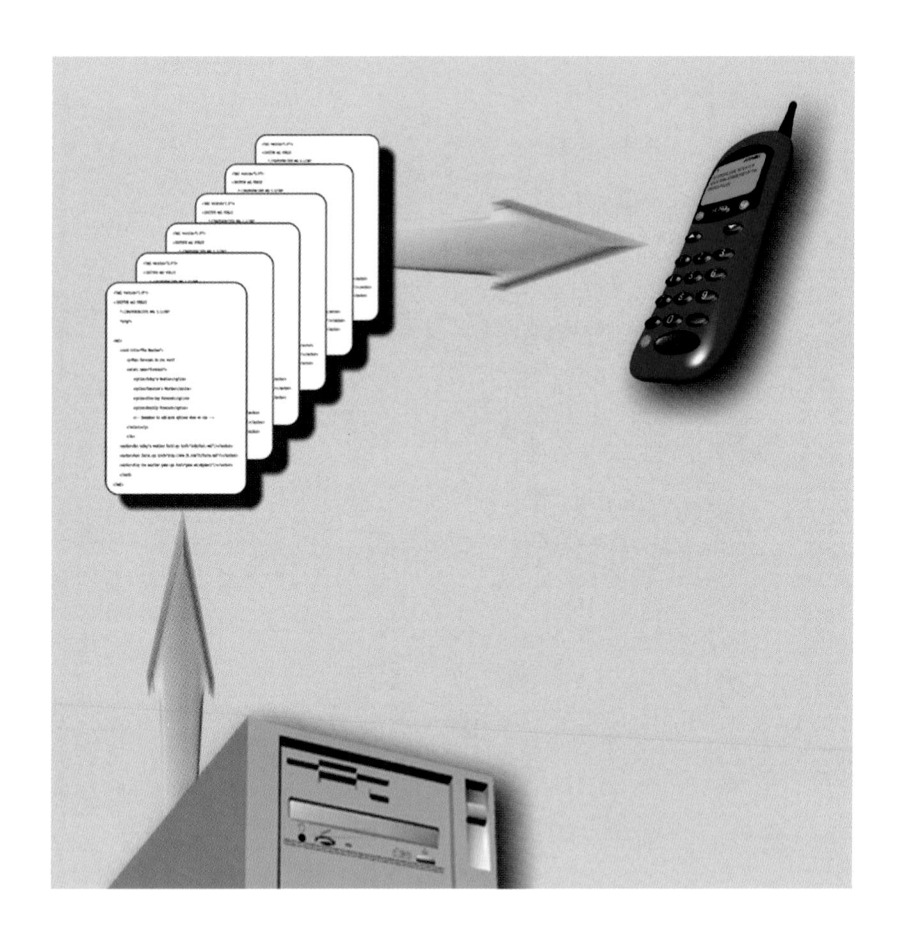

3G A standard for the next generation (third generation) of cell phones, which will be able to access the Internet at high data rates.

802.11 The most common standard for wireless computing networks. Several 802.11 standards allow for different rates of transmission.

advanced mobile phone network (AMPS) The first generation of cellular networks; they are analog-based.

amplifier A device that strengthens a signal by increasing its amplitude.

amplitude The magnitude of a wave.

amplitude modulation (AM) A method of modulation in which an information signal is superimposed over a carrier signal by varying the amplitude of the carrier signal.

analog data Information represented as a continuous wave in which there can be infinite variations between two points.

analog-to-digital converter (ADC) A device that converts analog data to digital data.

antenna A device that sends and receives radio signals by converting alternating voltages to and from electromagnetic fields.

antenna gain A measure of an antenna's capability to concentrate or receive electromagnetic energy in or from a given direction.

base service set (BSS) In an 802.11 network, an access point, along with all the wireless clients, such as computers and personal digital assistants, communicating with it.

base station A device in a cellular network that handles radio frequency communications with phones and other cellular devices inside a single cell.

base transceiver station See *base station*.

binary amplitude shift keying (BASK) A method of digital AM used to transmit digital data in some digital wireless systems.

Bluetooth A wireless networking standard that allows many different kinds of devices to communicate in a peer-to-peer fashion; that is, without having to use a server or other hardware to connect them.

card A single page on the Internet built with the WML language, designed to be viewed by cell phones.

Carrier Sense Multiple Access with Collision Avoidance (CSMA/CA) The way that computers communicate with a wireless access point. When used, transmitters/receivers transmit only when the channel is clear, thus avoiding "collisions," simultaneous transmissions that garble both transmitters' data.

carrier signal An RF wave used to carry information.

cell A geographic area in a cellular network that contains radio base stations, antennas, power sources, and communications to a central switching facility.

cellular network A network that uses a series of overlapping cells as a way to allow wireless devices within it to communicate.

client A piece of software running on a local computer or device that communicates with a central server.

cloning Copying information from a cell phone so that other people can make phone calls from a cell phone and charge them to the cloned phone.

Code Division Multiple Access (CDMA) A digital technique that allows several cell phones to share the same channel simultaneously by assigning each phone its own code, and having each phone receive coded messages at the same time.

Compact HTML (cHTML) A markup language similar to HTML used to build Internet sites for the i-mode service.

control channel A cellular communications channel that transmits system information and gives instructions to cell phones.

deck A group of related cards on the Internet built using the Wireless Markup Language, designed to be viewed by cell phones.

decryption A method of unscrambling encrypted data so that it can be understood.

demodulation The act of separating information from a carrier wave.

demodulator A device that separates information from a carrier wave.

digital data Data represented as bits that are either on or off. All data in computers is digital data.

digital signal processor A programmable chip that processes signals in a variety of ways so they can be sent or decoded more easily.

digital TV (DTV) A technique in which TV broadcasters and receivers use digital technologies. Digital TV offers a much higher resolution than analog television, and also allows for extra interactive features.

digital-to-analog converter (DAC) A device that converts digital data to analog data.

directional antenna An antenna that transmits in a single direction.

electromagnetic radiation Waves of energy of varying wavelengths and frequencies propagated through space.

electromagnetic spectrum The entire range of wavelengths or frequencies of electromagnetic radiation, such as visible light, the radio frequency, X-rays, infrared, and so on.

electromagnetic waves See *electromagnetic radiation.*

electronic serial number (ESN) A cell phone's serial number, programmed into the phone's number assignment module, that uniquely identifies the cell phone to the cellular system. It also helps guard against cell phone fraud.

encryption A method of scrambling data so it can be read only by its intended recipient.

Ethernet The most common local area networking standard.

eXtensible Markup Language (XML) An extension of HTML that separates the content of a Web page from its display. It can be used to allow designers to easily create Web pages to be displayed on many different devices, such as computers, cell phones, and PDAs.

extremely high frequency (EHF) Electromagnetic waves between 30 and 300 GHz; used for satellite transmissions and for radar.

extremely low frequency (ELF) Electromagnetic waves below 3 kHz; used for submarine communications.

family radio service (FRS) A walkie-talkie type of radio with extra features that allows people to easily talk with each other within an area of several miles.

Federal Communications Commission (FCC) The government agency that regulates the airwaves.

filter A device, often used in a receiver or tuner, that discards all signals except select ones.

firewall A hardware or hardware/software combination that protects computers on a network from being attacked by hackers or snoopers.

frequency The number of times per second that wave cycles occur.

frequency modulation (FM) A method of modulation in which an information signal is superimposed over a carrier signal by varying the frequency of the carrier signal.

frequency reuse A technique that allows cell phone networks to use the same frequency for different subscribers in different cells.

gateway mobile switching center (GMSC) A center that routes calls to and from a cellular network to the public phone system and other cellular networks.

geostationary satellite A satellite that orbits at the same speed as the earth so it stays over the same location on earth.

gigahertz One billion hertz; one billion cycles per second.

global positioning system (GPS) A system that allows you to pinpoint your location on earth using satellites.

Global System for Mobile Communications (GSM) A standard for digital cellular communications developed in Europe that allows European countries to have a single cellular standard. It uses TDMA as its way of communicating, and operates in different frequencies in different countries.

handoff A technique in which, when a cell phone subscriber travels from one cell to another, the communications with the network are transferred from one base station to another.

hertz A measurement of frequency that equals one cycle per second.

high frequency (HF) Another term for short wave. See *short wave*.

High-Definition TV (HDTV) The highest resolution of digital TV. It includes high-quality Dolby Digital surround sound.

home location register (HLR) The database in a cellular network that keeps tracks of all subscribers' current locations in the network.

HTTP (Hypertext Transfer Protocol) An Internet protocol that defines the way Web browsers and Web servers communicate with each other.

hub A device that connects several computers to one another on a network.

hub/router A combination of a hub and router that connects computers, routes data among them, and provides access to the Internet or other networks. Home networks commonly use a hub/router.

Hypertext Markup Language (HTML) The language used to build Web pages.

i-mode A way of sending data and interactive services over the Internet through cell phones; used primarily in Japan.

infrared port A port on a computer or other device through which infrared signals are sent.

Internet service provider (ISP) A company that provides Internet access to people for a fee.

IrDA (Infrared Data Association) A standard for using an infrared port on a computer or other device for communications.

IP address An Internet address, such as 126.168.5.22, that computers need to get onto the Internet.

kilohertz One thousand hertz; one thousand cycles per second.

landline A telephone line that uses wires.

line of sight A method of transmission in which the sending and receiving devices must be in a line with each other, with no obstacles between them.

local area network (LAN) A network that allows computers to send and receive information among each other, and to do other communications tasks.

low frequency or long wave (LW) Electromagnetic waves between 20 and 300 kHz; used in AM radio broadcasting.

low-earth orbit (LEO) satellites Satellites that orbit at a height of 100 kilometers to 1,000 kilometers. They can be used for satellite phones as well as for various sensing applications, telemetry, navigation, and spying.

low-noise amplifier An amplifier that amplifies very weak signals; often used in receivers.

low-power FM radio (LPFM) A method of FM radio broadcasting that allows nonprofit groups to broadcast to a small geographic area, such as a neighborhood, city, or town.

mail server A server that delivers or receives e-mail.

mCommerce (mobile commerce) Using a cell phone or other cellular device to shop or do other kinds of commerce.

medium frequency (MF) or medium wave (MW) Electromagnetic waves between 300 and 3000 kHz; used in AM radio broadcasting.

megahertz One million hertz; one million cycles per second.

microbrowser A browser that a cell phone or similar devices uses to browse the World Wide Web.

microwaves Electromagnetic waves in the UHF, SHF, and EHF spectrum. They have the highest frequencies in the RF band and, because of that, they have the smallest wavelengths.

middle-earth orbit (MEO) satellites Satellites that orbit between 5,000 and 15,000 kilometers above the earth. They can be used in the global positioning system (GPS).

mixer A device that combines signals as a way to separate information from carrier waves or to add information to carrier waves.

mobile electronic identity number (MEIN) A serial number that identifies someone in a GSM system. It is programmed into a SID card.

mobile electronic transactions (MET) A standard for mCommerce that includes encryption and other ways of protecting people's privacy and data.

mobile identification number (MIN) A number, programmed into a cell phone's number assignment module, that identifies a cell phone subscriber.

mobile subscriber unit (MSU) or mobile system (MS) Another term for a cell phone.

mobile switching center (MSC) The "brains" of a cellular network; it handles the processing and routing of cell calls. Each MSC is in charge of several cells and base stations. They sometimes are referred to as mobile telephone switching office (MTSO), mobile-service switching center (MSC), or mobile telephone exchange (MTX).

modulation The process or technique of modifying waves to transmit information.

modulator A device that modifies carrier waves to transmit information.

MPEG-2 A method of compressing digital animation and TV signals that reduces their size but still retains their high quality. Digital TV and HDTV use MPEG-2.

network address translation (NAT) A technique in a local area network that provides an internal IP address to computers inside the network, while masking the IP address to the outside world.

network card An add-in card put into a computer so it can get onto a network.

number assignment module (NAM) Internal memory in a cell phone that has programmed into it identifying information about the phone, including the mobile identification number, the system ID, and the features that a subscriber has paid for.

omnidirectional antenna An antenna that transmits or receives equally well in all directions.

oscillator A device that creates a wave at a specific wavelength.

overhead signal A communications channel in a cellular network that contains identifying information about the network, as well as commands to cell phones.

packet Data that has been broken down into pieces for transmission over the Internet or another network.

palm query application (PDA) A small piece of software on a wireless Palm device that allows it to get information from the Internet using Web clipping.

palmtop computer A small computer, such as the Palm, that fits in the palm of your hand, and often is used for keeping track of schedules, to-do lists, and a calendar. It also can be used for wireless communications.

PCMCIA A standard for allowing laptop computers to use add-in cards such as network cards.

PCS (personal communications services) A digital cellular network operating in the U.S. in the 1900 Mhz band. It offers a variety of communications services that analog systems can't offer.

peer-to-peer network A network that allows computers or other devices to connect directly with one another without having to use a server or other hardware to connect them.

personal digital assistant (PDA) A small handheld computer, such as a Palm device or Windows CE device.

petahertz One quadrillion hertz; one trillion cycles per second.

phase modulation (PM) A variant of FM. It's useful for sending digital data over cellular networks, in which the phase of a wave is continually shifted as a means of modulation.

piconet A network formed by the connection of two or more Bluetooth devices with one another.

POP3 (Post Office 3) An Internet communications standard used to receive e-mail.

propagation loss The weakening of a signal as it travels through the atmosphere.

radio frequency (RF) The portion of the electromagnetic spectrum used to transmit information.

receiver A device that receives information from an antenna and processes the information so that it can be used in some way.

router A piece of hardware that sends data to its proper destination on the Internet or on a local area network. Routers work by examining the destination address of each piece of data and sending it toward its final destination.

server A computer, especially on the Internet, that performs some task for other computers, such as sending or receiving e-mail or delivering Web pages.

short message service (SMS) A service that allows people with cell phones to send text messages to each other.

short wave (SW) Electromagnetic waves between 3 and 30 MHz; used in AM broadcasting and in shortwave and amateur radio.

signal processing The act of manipulating a signal to make it transmit more effectively, or, after it's received, to be understood by a device more effectively.

Simple Mail Transfer Protocol (SMTP) An Internet communications standard used to send e-mail.

subscriber identity card (SIM) In a GSM system, a card that identifies a cell phone user and allows him to use other people's cell phones, in other countries, while billing him for his use.

subscription satellite radio A business in which many high-quality radio broadcasts are delivered by satellite to subscribers for a monthly fee.

super-high frequency (SHF) Electromagnetic waves between 3 and 30 GHz; used in fixed wireless communications and for satellite transmissions.

system ID (SID) A number that identifies a cellular network. It is programmed into a cell phone's number assignment module.

TCP/IP (Transmission Control Protocol/Internet Protocol) The communications protocols that underlie the Internet.

terahertz One trillion hertz; one trillion cycles per second.

Time Division Multiple Access (TDMA) A digital technique that allows several cell phones to use the same channel simultaneously by giving each phone its own dedicated time slot in the channel.

transceiver A device that includes both a transmitter and a receiver.

transmitter A device that sends RF signals carrying information.

two-way pager A pager that allows someone to both send and receive messages and pages. Some two-way pagers are used to receive and send e-mail.

UART (Universal Asynchronous Receiver and Transmitter) chip A chip that handles communications in computers and palmtop computers.

ultra high frequency (UHF) Electromagnetic waves between 300 and 3,000 MHz; used in television broadcasting and by cellular telephones.

uniform resource locator (URL) An address on the Internet, such as `www.zdnet.com`, that allows computers and other devices to visit it.

very high frequency (VHF) waves Electromagnetic waves between 30 and 300 MHz; used in FM radio and television broadcasting.

very low frequency (VLF) waves Electromagnetic waves between 3 and 30 kHz; used in maritime communications.

virtual private network (VPN) An encryption technique that allows people to connect to their corporation's network over the Internet, while protecting the data from being seen by anyone else.

voice channel The channel in a cellular network used for transmitting voice signals.

voice coding The compression of a digital voice signal so that it can be transmitted using less bandwidth than if it weren't compressed.

Voice eXtensible Markup Language (VXML) An extension of XML that allows people to get information from, and interact with, the Internet, by using their voice.

WAP Transaction Protocol (WTP) A communications protocol, part of the Wireless Access Protocol (WAP), that is the equivalent of the Internet's TCP/IP protocols. It allows cell phones and similar devices to access the Internet.

wavelength The length of an electromagnetic wave; in other words, the length between the wave's peaks.

Web browser A piece of software that allows people to browse the World Wide Web.

Web clipping A technique that allows Palm devices to get information from the Internet.

whip antenna A kind of antenna often used in automobiles.

wireless access point A device that connects wireless devices, such as a computer equipped with a wireless network card, to a network.

Wireless Access Protocol (WAP) An Internet protocol that defines the way in which cell phones and similar devices can access the Internet.

wireless bridge See *wireless point-to-point networks*.

Wireless Markup Language (WML) A markup language related to HTML that is used to create Web sites that cell phones and similar devices can visit.

wireless network A network of computers, phones, or other devices that can communicate without wires.

wireless point-to-point networks Networks that use fixed transmitters and receivers with a clear line of sight between them to send and receive network communications. It allows companies with more than one building to extend their networks across the buildings. Also called a wireless bridge.

Wireless Transport Layer Security (WTLS) A communications protocol that allows cellular phones to send and receive encrypted information over the Internet.

WMLScript A scripting language that allows for interaction between the cell phone and the Internet.

Yagi antenna A kind of antenna often used for TV reception and amateur radio. It's also known as the Yagi-Uda antenna. It was designed to provide high gain for VHF and UHF RF signals.

Index

X-Y-Z